Biodiversity

Global Patterns, Conservation Strategies, and Ecosystem Resilience

nova science publishers
www.novapublishers.com
New York

Conservation Biology and Biodiversity

More information about this series can be found at

Environmental Research Advances

More information about this series can be found at

Biodiversity

Global Patterns, Conservation Strategies, and Ecosystem Resilience

nova
science publishers
www.novapublishers.com
New York

Conservation Biology and Biodiversity

More information about this series can be found at

Environmental Research Advances

More information about this series can be found at

Mohammad Basyuni
Faujiah Nurhasanah Ritonga
Editors

Biodiversity

Global Patterns, Conservation Strategies, and Ecosystem Resilience

nova
science publishers
www.novapublishers.com
New York

DOI: https://doi.org/10.52305/LKIC0872

Additional color graphics may be available in the e-book version of this book.

Library of Congress Cataloging-in-Publication Data

ISBN: 979-8-90134-246-6 (Softcover)
ISBN: 979-8-90134-310-4 (eBook)

Published by Nova Science Publishers, Inc. † New York

Contents

Preface

Life on Earth is a masterpiece of variety and connection. From the soaring canopies of tropical rainforests to the invisible microbes weaving chemical cycles beneath mangrove mud, biodiversity shapes every breath of air we inhale, every meal we eat, and every drop of water we drink. Biodiversity underpins ecosystem functioning and human well-being, yet global assessments consistently document rapid declines across terrestrial, freshwater, and marine realms. *Biodiversity: Global Patterns, Conservation Strategies, and Ecosystem Resilience* presents eight chapters that examine these challenges through integrated scientific and management perspectives. The book takes readers on a compelling journey through this unfolding drama, revealing the science, the stories, and the solutions that can still secure a thriving planet.

The book opens with an inspiring look at the tropical forests of Sumatra to discover the remarkable resilience of Dipterocarpaceae trees, keystone giants struggling to recover from deforestation. The journey returns to Sumatra to examine the biodiversity and ecosystem functions of its mangroves, ecosystems that shield coasts, store vast amounts of carbon, and sustain fisheries. Turning to Indonesia's coastlines, readers are guided through a practical and fascinating comparison of mangrove sampling methods, where science meets fieldwork to improve how we measure and monitor nature's richness. Beneath these mangrove canopies lies an invisible world: the next chapter uncovers the microbial engines that regulate greenhouse gas emissions, revealing how microscopic life helps stabilize the planet's climate.

From there, the narrative dives into the turquoise waters of the Coral Triangle to explore coral reefs battling nutrient stress, then follows the shifting currents of Saleh Bay, where oceanographic changes dictate the fortunes of pelagic fish and the communities who depend on them. Then the volume continued with an exploration of faith-based and localized conservation frameworks, demonstrating how cultural values and governance innovations can complement international biodiversity targets.

The closing synthesis integrates these regional and thematic studies into a global framework for biodiversity conservation, highlighting scalable strategies to enhance ecosystem resilience, link oceanographic variability to pelagic fish productivity in Saleh Bay. By combining case studies, methodological advances, and cross-scale syntheses, this book offers researchers, conservation practitioners, and policymakers a concise scientific resource for understanding biodiversity dynamics and informing effective, evidence-based conservation action.

We are grateful to the following reviewers (Dr. Chen Su, Northeast Forestry University, Dr. Gao Jianwe, Institute of Vegetables, Shandong Academy of Agricultural Sciences, Dr. Itchika Sivaipram, Chulalongkorn University, Dr. Siti Halimah Larekeng, Universitas Hasanuddin, and Dr. Rinny Rahmania, Badan Riset and Inovasi Nasional), whose insightful comments and constructive suggestions greatly enhanced the clarity, scientific rigor, and overall quality of this volume. Their careful reading and thoughtful feedback strengthened each chapter and ensured that the book meets the highest standards of scientific publication and relevance.

Biodiversity: Global Patterns, Conservation Strategies, and Ecosystem Resilience is not just a reference,it is an invitation to see the planet as it truly is: alive, interconnected, and worth defending.

Mohammad Basyuni
Faujiah Nurhasanah Ritonga
Editors
Center of Excellence for Mangrove
Universitas Sumatera Utara, Medan, Indonesia

Chapter 1

The Resilience of Dipterocarpaceae Trees to Disturbance on Sumatra Island, Indonesia

Faujiah Nurhasanah Ritonga[1,2,*]
Mohammad Basyuni[1,2]
Chen Su[3]
and Song Runxian[4]

[1]Center of Excellence for Mangrove, Universitas Sumatera Utara, Medan, Indonesia
[2]Faculty of Forestry, Universitas Sumatera Utara, Pancurbatu, Deli Serdang, Indonesia
[3]Key Laboratory of Tree Genetics and Breeding, Northeast Forestry University, Harbin, China
[4]Center for Translational Medicine, Precision Medicine Institute, The First Affiliated Hospital, Sun Yat-sen University, Guangzhou, China

Abstract

Dipterocarpaceae, a dominant tree family in the tropical forests of Southeast Asia, is both ecologically crucial and economically valuable, particularly on Sumatra Island, Indonesia. Despite their role as keystone species that support biodiversity and ecosystem stability, many dipterocarp species face severe threats from deforestation and land conversion, especially oil palm plantations, and overexploitation. This chapter evaluates their ecological resilience to disturbance, defined as their ability to survive, regenerate, and restore ecosystem functions after environmental stress. The findings show that while some dipterocarps demonstrate recovery potential under favorable conditions, many remain highly vulnerable due to regeneration bottlenecks, limited seed dispersal, and ongoing habitat fragmentation. Current conservation efforts,

* Corresponding Author's Email: faujiahnurhasanah@usu.ac.id

In: Biodiversity
Editors: Mohammad Basyuni and Faujiah Nurhasanah Ritonga
ISBN: 979-8-90134-246-6

including protected areas, reforestation, and community-based management, have had mixed success, often hindered by weak governance and resource pressures. The strengthening of landscape-scale restoration, the enforcement of legal protection, and the integration of climate adaptation into management strategies are urgently needed to ensure the long-term survival of Dipterocarpaceae in forests in Sumatra.

Keywords: Dipterocarpaceae conservation, ecological resilience, forest fragmentation, Sumatra forests

Introduction

The Dipterocarpaceae family forms the ecological backbone of Southeast Asia's tropical rainforests, dominating lowland and hill forest canopies and supporting critical ecosystem functions such as carbon storage, nutrient cycling, and biodiversity conservation (Basyuni et al., 2019; Cvetković et al., 2022). On Sumatra, dipterocarps also serve as a key source of high-value timber, making them central to both forest conservation and exploitation (Latifah & Zahrah, 2021; Ritonga et al., 2018). Owing to their dual ecological and economic roles, understanding their long-term survival and management is a pressing priority.

Despite their ecological and economic importance, dipterocarps remain understudied at the species level in terms of their resilience to disturbance. Most research has focused on forest-level recovery, often overlooking species-specific traits and regeneration strategies (Inada et al., 2017). Addressing this gap is essential for predicting species persistence and ensuring effective forest restoration efforts.

Sumatra has long been a hotspot of land-use change, experiencing extensive forest conversion due to logging, oil palm and pulpwood plantations, mining, and infrastructure expansion (Aso et al., 2024). These pressures fragment habitats, disrupt ecological processes such as pollination and seed dispersal, and alter microclimates, posing particular challenges for dipterocarps that rely on mast fruiting and specialized dispersal mechanisms (Adrianto et al., 2019; Visser et al., 2011). The cumulative effects of these disturbances threaten not only dipterocarp populations but also the ecological integrity of entire landscapes.

Resilience, defined as a forest's ability to recover while maintaining essential structures and functions, is shaped by both species traits and environmental factors such as landscape connectivity, soil quality, and microclimate (Chambers et al., 2019; Sánchez-Pinillos et al., 2024). Understanding these dynamics at the species level can guide targeted conservation, restoration efforts, and sustainable management strategies. A species-focused approach offers a clearer picture of ecosystem responses and better informs conservation priorities.

This chapter aims to assess the resilience of Dipterocarpaceae trees in Sumatra under anthropogenic disturbance. This section reviews the ecological characteristics of dipterocarps, the extent and drivers of forest disturbance, and resilience indicators at the species and ecosystem levels. By synthesizing current knowledge and case studies, this chapter contributes to evidence-based strategies for conservation, restoration, and forest governance.

Historical and Contemporary Dipterocarpaceae in Sumatera

Historically, Sumatra forests have been among the most species-rich and structurally complex tropical ecosystems in the world, with Dipterocarpaceae dominating lowland and hill forest canopies (Purwaningsih, 2004). Vast, contiguous forest tracts supported exceptional biodiversity, stabilized local climates, and sustained forest-dependent communities. Dipterocarps such as Shorea, Dipterocarpus, Dryobalanops, and Hopea species tower over the canopy, some exceeding 60 m, exhibit unique ecological traits such as slow growth and synchronized mast fruiting every few years (Bartholomew et al., 2022; Visser et al., 2011). These characteristics allow keystone species to maintain forest dynamics and regeneration.

Prior to the middle 20th century, large areas of Sumatra's forests remained intact, with minimal selective logging and traditional forest management practices by local communities that preserved natural regeneration cycles (Susilowati et al., 2024; Tito de Morais et al., 2015). However, the onset of industrial-scale timber extraction, plantation development (especially oil palm and pulpwood), and infrastructure expansion in the latter half of the century rapidly altered these landscapes. By the early 2000s, Sumatra had lost more than half its natural forest cover, leaving dipterocarps particularly vulnerable due to their shade tolerance, slow growth, and reliance on undisturbed habitats (Hardiyanto et al., 2024; Mambrasar et al., 2019).

Table 1. Some of the most prominent genera and representative dipterocarp species in Sumatera

No	Species	Genus	Category of IUCN Red List	Habitat	References
1	*Anisoptera costafa*	Anisoptera	Vulnerable	Pasir Mayang, Jambi	(Affand et al., 1999)
2	*Anisoptera marginata*		Vulnerable	Pasir Mayang, Jambi	(Affand et al., 1999)
3	*Cotylelobium melanoxylon*	Cotylelobium	Endangered	Mursala Island, Central Tapanuli	(Iswanto et al., 2020; Robiansyah et al., 2020; Susilowati et al., 2019)
4	*Dipterocarpus cinereus*	Dipterocarpus	Critically Endangered	Mursala Island, Central Tapanuli	(Rachmat & Subiakto, 2015; Robiansyah et al., 2020)
5	*Dipterocarpus cornutus*		Critically Endangered	Mursala Island, Central Tapanuli	(Robiansyah et al., 2020)
6	*Dipterocarpus gracilis*		Vulnerable	Mursala Island, Central Tapanuli	(Robiansyah et al., 2020)
7	*Dipterocarpus kunstleri*		Critically Endangered	Mursala Island, Central Tapanuli	(Robiansyah et al., 2020)
8	*Dryobalanops aromatica*	Dryobalanops	Critically Endangered	Central Tapanuli; Aceh Singkil; Subulussalam; Pakpak Barat; West Sumatera	(Aswandi & Kholibrina, 2021; Ritonga et al., 2018)
9	*Dryobalanops oblongifolia*			Tanjung Jabung, Jambi	(Siregar et al., 2019)
10	*Dryobalanops sumatrensis*		Vulnerable	Mursala Island, Central Tapanuli; Tanjung Jabung, Jambi	(Robiansyah et al., 2020; Siregar et al., 2019)
11	*Hopea bancana*	Hopea	Critically Endangered	Mursala Island, Central Tapanuli	(Robiansyah et al., 2020)
12	*Hopea mengerawan*		Critically Endangered	Mursala Island, Central Tapanuli	(Robiansyah et al., 2020)
13	*Shorea acuminata*	Shorea	Critically Endangered	Mursala Island, Central Tapanuli	(Robiansyah et al., 2020)
14	*Shorea atrinervosa*		Vulnerable	Mursala Island, Central Tapanuli	(Robiansyah et al., 2020)
15	*Shorea bracteolata*		Endangered	Mursala Island, Central Tapanuli	(Robiansyah et al., 2020)
16	*Shorea furfuracea*		Vulnerable	Mursala Island, Central Tapanuli	(Robiansyah et al., 2020)
17	*Shorea gibosa*		Vulnerable	Pasir Mayang, Jambi	(Affand et al., 1999)
18	*Shorea guiso*		Vulnerable	Pasir Mayang, Jambi	(Affand et al., 1999)

No	Species		Category of IUCN Red List	Habitat	References
19	*Shorea johorensis*		Critically Endangered	Mursala Island, Central Tapanuli	(Robiansyah et al., 2020)
20	*Shorea lepidota*		Critically Endangered	Mursala Island, Central Tapanuli	(Robiansyah et al., 2020)
21	*Shorea leprosula*		Near Threatened	Mursala Island, Central Tapanuli	(Robiansyah et al., 2020)
22	*Shorea multiflora*		Least Concerned	Mursala Island, Central Tapanuli	(Robiansyah et al., 2020)
23	*Shorea ochrophloia*		Critically Endangered	Mursala Island, Central Tapanuli	(Robiansyah et al., 2020)
24	*Shorea ovalis*		Least Concerned	Pasir Mayang, Jambi	(Affand et al., 1999)
25	*Shorea ovata*		Endangered	Mursala Island, Central Tapanuli	(Robiansyah et al., 2020)
26	*Shorea palembanica*		Critically Endangered	Mursala Island, Central Tapanuli	(Robiansyah et al., 2020)
27	*Shorea parvifolia*		Least Concerned	Mursala Island, Central Tapanuli	(Robiansyah et al., 2020)
28	*Shorea pinanga*		Least Concerned	Pasir Mayang, Jambi	(Affand et al., 1999)
29	*Shorea retinodes*		Vulnerable	Mursala Island, Central Tapanuli	(Robiansyah et al., 2020)
30	*Shorea selanica*		Critically Endangered	Pasir Mayang, Jambi	(Affand et al., 1999)
32	*Shorea seminis*		Vulnerable	Pasir Mayang, Jambi	(Affand et al., 1999)
33	*Vatica pauciflora*	Vatica	Vulnerable	Mursala Island, Central Tapanuli	(Robiansyah et al., 2020)
34	*Vatica perakensis*		Endangered	Mursala Island, Central Tapanuli	(Robiansyah et al., 2020)
35	*Vatica soepadmoi*		Critically Endangered	Mursala Island, Central Tapanuli	(Robiansyah et al., 2020)
36	*Vatica stapfiana*		Vulnerable	Mursala Island, Central Tapanuli	(Robiansyah et al., 2020)
37	*Vatica venulosa ssp. Simalurensis*		Critically Endangered	Mursala Island, Central Tapanuli	(Robiansyah et al., 2020)

A previous study revealed these dramatic shifts, showing a decline in natural forests from 78.3% cover in 1950 to just 46.8% cover in 2017, with lowland dipterocarp-rich forests suffering the greatest losses (Santoro et al., 2023). Although these maps do not isolate dipterocarp distributions, their historical dominance in lowland forests strongly indicates severe range contraction, especially in regions where deforestation exceeded 50% of the land area.

Today, the remaining dipterocarp-rich forests are often restricted to protected areas or remote uplands, whereas lowland forests are heavily fragmented and degraded (Nguyen et al., 2022). In logged or secondary forests, natural regeneration is constrained by reduced seed sources, soil compaction, invasive species, and altered microclimates, leading to low seedling recruitment and disrupted forest structure (Fitrisia & Asri, 2021).

Dipterocarp resilience varies widely. Some species can recover in selectively logged areas or enriched planting sites, yet many struggle to regenerate in degraded landscapes because of dispersal limitations, fire risk, and edge effects (Seidl et al., 2024; Sorenson et al., 2025). Active restoration approaches, such as the cultivation of *Shorea leprosula* or *Dryobalanops aromatica*, have shown promise but require intensive maintenance and long-term monitoring (Attarik et al., 2024; Sabrina & Sinaga, 2023). Over time, secondary forests often shift toward pioneer species dominance, indicating a decline in Dipterocarp ecological roles (Ådjers et al., 1995).

Sumatra hosts more than 60 Dipterocarp species, many of which are now classified as Vulnerable, Endangered, or Critically Endangered under the IUCN Red List (Table 1). This trend underscores the urgent need for targeted conservation and restoration strategies that address species-specific vulnerabilities and habitat requirements (IUCN, 2023).

Scientific Studies on Dipterocarpaceae and Forest Disturbance in Sumatra

1. Early Studies: Taxonomy, Ecology, and Timber Potential

The scientific exploration of Dipterocarpaceae in Sumatra began in the late 19th and early 20th centuries and was largely driven by colonial-era forestry initiatives (Torquebiau, 1984). Early botanists and foresters cataloged the rich diversity of dipterocarp species, documenting their dominance in lowland forests and their value as timber resources (Affand et al., 1999; Cao et al., 2009; Nguyen et al., 2022). These foundational studies emphasized species identification, wood properties, and basic ecological roles (Appanah & Turnbull, 1998; Cao et al., 2009; Kartawinata et al., 2008), laying the groundwork for future ecological and conservation research.

2. Midcentury and Industrial Expansion: Impacts of Logging and Fragmentation

By the mid-20th century, research shifted toward forest dynamics and structure, highlighting Dipterocarps' role in maintaining biomass and canopy stability (Harnelly et al., 2020). At this time, resilience was assumed rather than explicitly studied, as large-scale exploitation was still emerging. However, with the rise of industrial logging and land conversion in the latter half of the century, studies began documenting declines in dipterocarp regeneration and diversity. Research in logged forests revealed that seedling and sapling recruitment is often hindered by soil compaction, insufficient light control, and the removal of mature seed-bearing trees (Priatna et al., 2006; Romadini et al., 2022; Sukendro & Aisyiyah, 2023). Landscape fragmentation further compounded these issues, reducing connectivity and altering ecological processes such as pollination and seed dispersal (Ritonga et al., 2018; Susilawati et al., 2018).

3. Contemporary Research: Resilience, Restoration, and Climate Context

Over the last three decades, ecological research has increasingly focused on dipterocarp resilience under anthropogenic disturbance. Advances in remote sensing, landscape ecology, and restoration science have revealed that natural recolonization is slow or absent in many degraded areas without active intervention (Hayward et al., 2021). Dipterocarps are now recognized as critical to carbon storage, climate adaptation, and biodiversity restoration, prompting numerous enriched plants and assisting in natural regeneration projects (Pratiwi et al., 2024). Despite this progress, major gaps remain, including limited data on species-specific responses to disturbance, recovery trajectories under different management regimes, and the compounded effects of climate change and land-use shifts. Future research must also integrate socioecological perspectives, incorporating local knowledge, policy, and governance frameworks to ensure sustainable conservation outcomes.

Ongoing Conservation Efforts for Dipterocarpaceae in Sumatra

a) Protected areas and forest reserve systems

One of the primary strategies for conserving Dipterocarpaceae in Sumatra is the establishment of national parks and forest reserves, such as Gunung Leuser, Bukit Barisan Selatan, and Kerinci Seblat National Parks. These areas safeguard extensive tracts of dipterocarp-rich lowland and montane forests, functioning as biodiversity strongholds and ecological corridors for large fauna and associated plant communities (Harapan et al., 2022; Essy Harnelly et al., 2018; Hartiningtias et al., 2019). However, while these protected zones have slowed deforestation, enforcement remains inconsistent. Data from recent park management reports indicate that illegal logging and agricultural encroachment continue, with some protected forests losing up to 1–2% of cover annually. This reveals a key gap between formal designation and on-the-ground protection, emphasizing the need for better surveillance systems and stronger engagement of local communities in park management.

b) Restoration and Reforestation Initiatives

In response to large-scale deforestation, various reforestation and restoration projects have been launched across Sumatra. Both government programs such as the Indonesian Ministry of Environment and Forestry's *Rehabilitation of Critical Land Program* and NGO-led projects focus on restoring degraded forests, often using native Dipterocarp species (Nawir et al., 2007). The enrichment of plants with species such as *Shorea leprosula, Dryobalanops aromatica*, and *Dipterocarpus gracilis* has become a key strategy (Otsamo et al., 1996). These projects typically involve planting dipterocarp seedlings under degraded or secondary forest canopies to accelerate recovery (Naiem et al., 2014). Other models include assisted natural regeneration (ANR), which enhances the recovery of natural forests by managing competing vegetation and protecting naturally regenerating dipterocarps (Dugan et al., 2003).

Community-based forest management community participation has emerged as a critical factor in dipterocarp conservation, with schemes such as Hutan Kemasyarakatan (HKm) and Hutan Desa providing legal frameworks for sustainable forest management (Martapani et al., 2021). Case studies have shown that villages adopting selective harvesting, enrichment planting, and fire prevention have achieved measurable forest cover stabilization, with some areas reporting up to 15% increases in dipterocarp seedling density over five years (Cerullo & Edwards, 2019). These programs succeed where land tenure rights are secure and benefits are shared, but many initiatives remain vulnerable to market fluctuations, political shifts, and a lack of technical expertise. Strengthening partnerships between communities, local governments, and conservation NGOs could enhance these outcomes.

c) Conservation Through Research and Seed Banks

Universities and research institutions, including IPB University and Universitas Sumatera Utara, play an increasingly strategic role in dipterocarp conservation through long-term ecological monitoring, genetic diversity assessments, and silvicultural trials (Susilawati et al., 2018; Susilowati et al., 2024). Seed banks and nurseries have become vital for safeguarding genetic diversity and supporting restoration programs, with collections of Shorea and Dipterocarpus species now available for large-scale planting. However, maintaining viable seeds remains challenging because Dipterocarp seeds are recalcitrant, limiting storage duration and requiring continuous propagation. Research on seed storage physiology and vegetative propagation methods could address these barriers and ensure long-term restoration success.

d) Policy Frameworks and International Support

Indonesia's national commitment to forest conservation is framed by broader policies such as the Moratorium on New Forest Concessions, the REDD+ framework, and targets aligned with the UN Sustainable Development Goals (SDGs). These frameworks, supported by international organizations such as the UNDP, WWF, and GIZ, have encouraged more strategic approaches to forest management. The REDD+ mechanism, in particular, has supported forest conservation by creating financial incentives for reducing emissions from deforestation and degradation. Several REDD+ pilot projects in Sumatra have included dipterocarp-rich forests, although implementation challenges remain, particularly in monitoring, governance, and benefit distribution (Indrajaya et al., 2022; Murray et al., 2015).

Challenges and Opportunities Ahead

Despite ongoing conservation initiatives, Dipterocarp forests in Sumatra face persistent threats, including illegal logging, weak law enforcement, insufficient funding, and limited ecological data (Appanah & Turnbull, 1998; Ritonga et al., 2018). Climate change further complicates recovery by intensifying droughts, fires, and pest outbreaks. Strengthening partnerships among government agencies, local communities, and NGOs, integrating traditional ecological knowledge, and investing in long-term monitoring are crucial to reversing these trends. Greater public awareness of the ecological and cultural significance of dipterocarps may also help secure stronger conservation commitments.

National and International Legal Frameworks on Forest Disturbance

Indonesia has developed an extensive legal and policy framework to address forest disturbance and biodiversity loss, including protections relevant to dipterocarps. However, implementation remains uneven, and some laws lack clear connections to ground-level realities. Below is a summary of the most critical regulations and agreements shaping dipterocarp conservation:

Key National Policies

1. Forestry Law No. 41/1999 – Provides the foundation for forest classification, management zones (protection, conservation, production), and penalties for illegal activities.
2. Permanent Forest Moratorium (Presidential Instruction No. 5/2019) – Prohibits new permits in primary forests and peatlands, safeguarding biodiversity-rich areas, including Dipterocarp habitats.

Key International Commitments

1. REDD+ (reducing emissions from deforestation and forest degradation) connects climate mitigation with forest protection, with Sumatra's dipterocarp-rich forests prioritized for their high carbon stocks.

2. CITES (Convention on International Trade in Endangered Species) – Regulates the trade of several Dipterocarp species to prevent overexploitation.

Other policies, such as environmental protection laws, community forestry programs, and SDG integration, further support conservation but could be summarized in a policy table or appendix for brevity. Strengthening enforcement, clarifying land tenure, improving local engagement, and aligning international initiatives such as the UN-REDD and FCPF with local realities are essential to improving conservation outcomes.

Best Strategies for Dipterocarpaceae Conservation in the Current Context

Conserving Dipterocarpaceae in today's era of deforestation, climate change, and habitat degradation requires adaptive and forward-looking approaches. While earlier sections detail existing conservation efforts, this section focuses on new, actionable strategies to strengthen protection and ensure the long-term survival of these keystone trees.

1. Community-based forest management should be integrated with scientific restoration. Empowering local communities remains a foundational approach in Dipterocarpaceae conservation. Building upon the achievements of social forestry programs such as Hutan Kemasyarakatan and Hutan Desa, future strategies should prioritize capacity building in sustainable forest management and ecological restoration techniques (Santika et al., 2017). Securing land tenure rights is essential for fostering community ownership and long-term stewardship. Furthermore, integrating traditional ecological knowledge with scientific practices, particularly in the selection and nurturing of dipterocarp species, can enhance the effectiveness of restoration while respecting local cultural values.
2. Remote sensing (e.g., SIPHON) and drones are used to detect illegal logging in real time. Local communities should be supported with mobile apps for reporting violations and making forest data publicly accessible to increase transparency (Paneque-Gálvez et al., 2014).
3. ANR should be prioritized in areas with existing seed sources to accelerate recovery. The enrichment of Dipterocarp species, plus

investment in seed banks and nurseries, ensures genetic diversity and resilience (Rajapakshe et al., (2024)).

4. The moratorium policies to include secondary forests should be strengthened. Interagency coordination should be improved to avoid conflicting land-use permits. REDD+ and climate finance programs for community engagement should be leveraged, and certification schemes for sustainably harvested forest products should be promoted (Pham et al., 2022).
5. Long-term monitoring plots should be established to track forest health and climate impacts. This study designs restoration to increase landscape connectivity and support species migration. Collaborate with platforms such as UN-REDD and GEF to exchange expertise and curb illegal trade under CITES (Büntgen & Rees, 2023).

Conclusion and Future Perspectives on Dipterocarpaceae Conservation and Forest Disturbance Management in Sumatra

Dipterocarpaceae forests in Sumatra represent a cornerstone of biodiversity and ecosystem services, yet their resilience is increasingly threatened by deforestation, habitat fragmentation, and climate change. While these forests exhibit remarkable ecological adaptations, their slow growth rates, specialized seed dispersal, and reliance on stable environments make recovery challenging without intervention. Understanding species-level vulnerabilities and ecosystem dynamics is therefore crucial for guiding conservation and restoration efforts.

Key challenges include persistent land-use pressures, weak enforcement of conservation policies, and limited financial and technical resources. Climate change adds another layer of complexity, altering rainfall patterns, increasing drought frequency, and accelerating biodiversity loss. Socioeconomic drivers, including agricultural expansion and unsustainable logging, further exacerbate these threats, highlighting the urgent need for integrated solutions that balance ecological integrity with community livelihoods.

Future conservation priorities should focus on restoring ecological connectivity, scaling up assisted natural regeneration (ANR), safeguarding genetic diversity through seed banks and research-driven species selection. Leveraging technology such as remote sensing, artificial intelligence, and genomic tools will enhance monitoring and adaptive management, while participatory governance models can ensure inclusive decision-making. Sustained collaboration among local communities, policymakers, researchers, and international partners, coupled with strong legal frameworks and sustainable financing, will be essential to securing the long-term survival of Dipterocarpaceae forests in Sumatra.

Disclaimer

None.

References

Ådjers, G., Hadengganan, S., Kuusipalo, J., Nuryanto, K., & Vesa, L. (1995). Enrichment planting of dipterocarps in logged-over secondary forests: effect of width, direction and maintenance method of planting line on selected Shorea species. *Forest Ecology and Management, 73*(1), 259-270.

Adrianto, H. A., Spracklen, D. V., & Arnold, S. R. (2019). Relationship Between Fire and Forest Cover Loss in Riau Province, Indonesia Between 2001 and 2012. *Forests, 10* (10), 889.

Affand, H., Nuryadin, A., & Prayoga, S. B. (1999). Screening for Natural Products of Shorea Spp. and Anisoptera Spp. of the Family Dipterocarpaceae From Pasir Mayang, Jambi, Sumatera. *Biotropia* (12).

Appanah, S., & Turnbull, J. M. (1998). A review of dipterocarps: taxonomy, ecology, and silviculture.

Aso, O., Muhsandi, K., & Gustiawan, I. (2024). *The land use history and origin of land status oil palm plantations in South Sumatera.* Paper presented at the IOP Conference Series: Earth and Environmental Science.

Aswandi, A., & Kholibrina, C. (2021). *New insights into Sumatran camphor (Dryobalanops aromatica Gaertn) management and conservation in western coast Sumatra, Indonesia.* Paper presented at the IOP Conference Series: Earth and Environmental Science.

Attarik, N., Pamoengkas, P., Rachmat, H. H., & Susilowati, A. (2024). Potential Use of Shorea leprosula for Rehabilitation of Degraded Tropical Production Forest Ecosystems. *Jurnal Sylva Lestari, 12*(2), 519-531.

Bartholomew, D. C., Banin, L. F., Bittencourt, P. R. L., Suis, M. A. F., Mercado, L. M., Nilus, R., ... Rowland, L. (2022). Differential nutrient limitation and tree height control leaf physiology, supporting niche partitioning in tropical dipterocarp forests. *Functional Ecology, 36*(8), 2084-2103.

Basyuni, M., Slamet, B., Sulistiyono, N., Thoha, A., Bimantara, Y., & Widjaja, E. (2019). *Species composition and plant diversity of logged-over forest in Sikundur, Gunung Leuser National Park, North Sumatra.* Paper presented at the IOP Conference Series: Earth and Environmental Science.

Büntgen, U., & Rees, G. (2023). Global change research needs international collaboration. *Science of the Total Environment, 902*, 166054.

Cao, C.-P., Gailing, O., Siregar, I. Z., Siregar, U. J., & Finkeldey, R. (2009). Genetic variation in nine Shorea species (Dipterocarpaceae) in Indonesia revealed by AFLPs. *Tree Genetics & Genomes, 5*(3), 407-420.

Cerullo, G. R., & Edwards, D. P. (2019). Actively restoring resilience in selectively logged tropical forests. *Journal of Applied Ecology, 56*(1), 107-118.

Chambers, J. C., Allen, C. R., & Cushman, S. A. (2019). Operationalizing Ecological Resilience Concepts for Managing Species and Ecosystems at Risk. *Frontiers in Ecology and Evolution, Volume 7 - 2019*.

Cvetković, T., Hinsinger, D. D., Thomas, D. C., Wieringa, J. J., Velautham, E., & Strijk, J. S. (2022). Phylogenomics and a revised tribal classification of subfamily Dipterocarpoideae (Dipterocarpaceae). *Taxon, 71*(1), 85-102.

Dugan, P., Durst, P. B., Ganz, D. J., & PJ, M. (2003). Advancing assisted natural regeneration (ANR) in Asia and the Pacific. *RAP publication, 19*.

Fitrisia, A., & Asri, Z. (2021). *Changing Function of the Forest and Exploitation in East Coast Sumatra in Colonial Period.* Paper presented at the 2nd *Progress in Social Science, Humanities and Education Research Symposium* (PSSHERS 2020).

Harapan, T. S., Nurainas, N., Syamsuardi, S., & Taufiq, A. (2022). Identifying the potential geographic distribution for *Castanopsis argentea* and *C. tungurrut* (Fagaceae) in the Sumatra Conservation Area Network, Indonesia. *Biodiversitas Journal of Biological Diversity, 23*(4).

Hardiyanto, E. B., Inail, M. A., Nambiar, S., & Mendham, D. S. (2024). Sustaining plantation forest productivity in Sumatra over three decades: From acacias to eucalypts. *Forest Ecology and Management, 553*, 121613.

Harnelly, E., Sara, A., Iqbar, Fathiya, N., & Umam, A. H. (2020). Relationship of Dipterocarpaceae in Ketambe Research Station, Gunung Leuser National Park based on morphological diversity of vegetative organs. IOP Conference Series: Earth and Environmental Science*, 425*.

Harnelly, E., Thomy, Z., & Fathiya, N. (2018). Phylogenetic analysis of Dipterocarpaceae in ketambe research station, gunung leuser national park (Sumatra, Indonesia) based on rbcL and matK genes. *Biodiversitas Journal of Biological Diversity, 19*(3), 1074-1080.

Hartiningtias, D., Andayani, N., & Winarni, N. L. (2019). Phenological pattern and community structure of dipterocarpaceae in Bukit Barisan Selatan National Park, Lampung. *Tropical Ecosystems: Structure, Functions and Challenges in the Face of Global Change*, 279-288.

Hayward, R. M., Banin, L. F., Burslem, D. F. R. P., Chapman, D. S., Philipson, C. D., Cutler, M. E. J., ... Dent, D. H. (2021). Three decades of postlogging tree community recovery in naturally regenerating and actively restored dipterocarp forest in Borneo. *Forest Ecology and Management, 488*, 119036.

Inada, T., Widiyatno, Hardiwitono, S., Purnomo, S., Putra, I. B. W., Kitajima, K., & Kanzaki, M. (2017). Dynamics Of Forest Regeneration Following Logging Management In A Bornean Lowland Dipterocarp Forest. *Journal of Tropical Forest Science, 29*(2), 185-197.

Indrajaya, Y., Yuwati, T. W., Lestari, S., Winarno, B., Narendra, B. H., Nugroho, H. Y. S. H., ... Mendham, D. (2022). Tropical Forest Landscape Restoration in Indonesia: A Review. *Land, 11*(3), 328.

Iswanto, A. H., Susilowati, A., Putra, A., Nopriandi, D., & Windra, E. (2020). *Natural durability of raru wood (Cotylelobium melanoxylon) against subterranean termite attack.* Paper presented at the *Journal of Physics: Conference Series.*

Report of the IUCN Species Survival Commission and Secretariat, (2023).

Kartawinata, K., Purwaningsih, P. T., Yusuf, R., Abdulhadi, R., & Riswan, S. (2008). Floristics and structure of a lowland dipterocarp forest at Wanariset Samboja, East Kalimantan, Indonesia. *Reinwardtia, 12*(4), 301-323.

Latifah, S., & Zahrah, M. (2021). *Potential spesies from the Dipterocarp family at Mandailing Natal Forest Production Management Unit, North Sumatra, Indonesia.* Paper presented at the IOP Conference Series: Earth and Environmental Science.

Mambrasar, Y., Damayanto, I., & Atikah, T. (2019). *Rediscovery Sumatran endemic flora: toward the establishment of data of biodiversity loss.* Paper presented at the IOP Conference Series: Earth and Environmental Science.

Martapani, A. N., Fauzi, H., & Naparin, M. (2021). The Impact of Community Forests on Social Economic Aspects Case Study: Sukamaju Peat Care Community KPH Kayu Tangi. *Jurnal Rimba Lestari, 1*(1), 36-46.

Murray, J. P., Grenyer, R., Wunder, S., Raes, N., & Jones, J. P. (2015). Spatial patterns of carbon, biodiversity, deforestation threat, and REDD+ projects in Indonesia. *Conservation Biology, 29*(5), 1434-1445.

Naiem, M., Widiyatno, & Al-Fauzi, M. Z. (2014). Progeny Test of Shorea Leprosula as Key Point to IncreaseProductivity of Secondary Forest in Pt Balik Papan Forest Industries, East Kalimantan, Indonesia. *Procedia Environmental Sciences, 20*, 816-822.

Nawir, A. A., Murniati, M., & Rumboko, L. (2007). *Forest rehabilitation in Indonesia: where to after more than three decades?*

Nguyen, C. H., Setyaningsih, C. A., Jahnk, S. L., Saad, A., Sabiham, S., & Behling, H. (2022). Forest Dynamics and Agroforestry History since AD 200 in the Highland of Sumatra, Indonesia. *Forests, 13*(9).

Otsamo, R., Otsamo, A., & Adjers, G. (1996). Reforestation experiences with dipterocarp species on grassland. Dipterocarp forest ecosystems: toward sustainable management. World Scientific Publishing, Singapore, 464-476.

Paneque-Gálvez, J., McCall, M. K., Napoletano, B. M., Wich, S. A., & Koh, L. P. (2014). Small Drones for Community-Based Forest Monitoring: An Assessment of Their Feasibility and Potential in Tropical Areas. *Forests, 5*(6), 1481-1507.

Pham, T. T., Vu, T. P., Hoang, T. L., Dao, T. L. C., Nguyen, D. T., Pham, D. C., ... Hoang, N. V. H. (2022). The Effectiveness of Financial Incentives for Addressing Mangrove Loss in Northern Vietnam. *Frontiers in Forests and Global Change, Volume 4 - 2021.*

Pratiwi, S., Pasaribu, N., & Siregar, J. P. (2024). Structure, Composition, Carbon Stocks of Dipterocarpaceae in West Batang Toru Forest Block, North Tapanuli Regency, North Sumatra. *Journal of Sylva Indonesiana, 7*(01), 22-29.

Priatna, D., Kartawinata, K., & Abdulhadi, R. (2006). Recovery of a lowland dipterocarp forest twenty two years after selective logging at Sekundur, Gunung Leuser National Park, North Sumatra, Indonesia. *Reinwardtia, 12*(3), 237-255.

Purwaningsih, P. (2004). Ecological distribution of Dipterocarpaceae species in Indonesia. *Biodiversitas Journal of Biological Diversity, 5*(2), 89-95.

Rachmat, H. H., & Subiakto, A. (2015). *Conserving the previously reported extinct tree species Dipterocarpus cinereus: An ex situ approach for the species conservation strategy.* Paper presented at the *Prosiding Seminar Nasional Masyarakat Biodiversitas Indonesia.*

Rajapakshe, R., Karthigesu, J., Thavananthan, S., Sivachandiran, S., Navaneetham, V., & Sinnamani, S. (2024). Restoring a dry tropical forest through assisted natural regeneration: enhancing tree diversity, structure, and carbon stock. *Trees, Forests and People, 17*, 100616.

Ritonga, F. N., Dwiyanti, F. G., Kusmana, C., Siregar, U. J., & Siregar, I. Z. (2018). Population genetics and ecology of Sumatran camphor (*Dryobalanops aromatica*) in natural and community-owned forests in Indonesia. *Biodiversitas Journal of Biological Diversity, 19*(6), 2175-2182.

Robiansyah, I., Hamidi, A., & Randi, A. (2020). High Species Diversity of the Family Dipterocarpaceae in Mursala Island, Indonesia. *International Journal on Advanced Science, Engineering and Information Technology, 10*(6), 2378-2385.

Romadini, N. P., Indrioko, S., WIDIYATNO, W., & Faridah, E. (2022). Regeneration failure and seedling growth of Dipterocarpus gracilis, a vulnerable dipterocarp in a tropical monsoon forest in Central Java, Indonesia. *Biodiversitas Journal of Biological Diversity, 23*(9).

Sabrina, R., & Sinaga, H. (2023). *The Conservation and Development Model Design of Camphor (Dryobalanapos Aromatica) Biodiversity in Sibagindar Champor Forest Pakpak Bharat District North Sumatra.* Paper presented at the *International Conference on Tropical Studies (ICTROPS).*

Sánchez-Pinillos, M., Dakos, V., & Kéfi, S. (2024). Ecological dynamic regimes: A key concept for assessing ecological resilience. *Biological Conservation, 289*, 110409.

Santika, T., Meijaard, E., Budiharta, S., Law, E. A., Kusworo, A., Hutabarat, J. A., ... Wilson, K. A. (2017). Community forest management in Indonesia: Avoided deforestation in the context of anthropogenic and climate complexities. *Global Environmental Change, 46*, 60-71.

Santoro, A., Piras, F., & Yu, Q. (2023). Spatial analysis of deforestation in Indonesia in the period 1950–2017 and the role of protected areas. *Biodiversity and Conservation.*

Seidl, R., Potterf, M., Müller, J., Turner, M. G., & Rammer, W. (2024). Patterns of early postdisturbance reorganization in Central European forests. *Proceedings of the Royal Society B: Biological Sciences, 291*(2031), 20240625.

Siregar, I., Kustiyarini, N., Wati, R., Rachmat, H., Siregar, U., & Dwiyanti, F. (2019). *Vegetative propagation of Dryobalanops sumatrensis and Dryobalanops oblongifolia subsp. oblongifolia by shoot cuttings.* Paper presented at the IOP Conference Series: Earth and Environmental Science.

Sorenson, Q. M., Young, D. J. N., & Latimer, A. M. (2025). Tree planting outcomes after severe wildfire depend on climate, competition, and priority. *Forest Ecology and Management, 575*, 122346.

Sukendro, A., & Aisyiyah, S. (2023). Systematic Review: History of Distribution and Conservation of Family Dipterocarpaceae through Vegetative Propagation. *Journal of Tropical Silviculture, 14*(02), 168-175.

Susilawati, A., Kholibrina, R. C., Rachmat, H. H., Elfiati, D., Aswandi, & Raeni, I. M. (2018). Macropropagation – An important tool for conservation of North Sumatran endangered tree species, *Dryobalanops aromatica. Biodiversitas Journal of Biological Diversity, 19*(5).

Susilowati, A., Rachmat, H. H., Elfiati, D., Kholibrina, C. R., Kusuma, Y. S., & Siregar, H. (2019). Population structure of Cotylelobium melanoxylon within vegetation community in Bona Lumban Forest, Central Tapanuli, North Sumatra, Indonesia. *Biodiversitas Journal of Biological Diversity, 20*(6).

Susilowati, A., Wijaya, K., Mawazin, M., Rachmat, H. H., Lismayati, Y., Kurniawan, H., & Ginting, I. M. (2024). Conserving dipterocarps biodiversity in remnant forests of small islands in Batam Island, Indonesia. *Biodiversitas Journal of Biological Diversity, 25*(2).

Tito de Morais, C., Ghazoul, J., Maycock, C. R., Bagchi, R., Burslem, D. F. R. P., Khoo, E., ... Kettle, C. J. (2015). Understanding local patterns of genetic diversity in dipterocarps using a multisite, multispecies approach: Implications for forest management and restoration. *Forest Ecology and Management, 356*, 153-165.

Torquebiau, E. (1984). Man-made dipterocarp forest in Sumatra. Agroforestry Systems, *2*(2), 103-127.

Visser, M. D., Jongejans, E., van Breugel, M., Zuidema, P. A., Chen, Y.-Y., Rahman Kassim, A., & de Kroon, H. (2011). Strict mast fruiting for a tropical dipterocarp tree: a demographic cost–benefit analysis of delayed reproduction and seed predation. *Journal of Ecology, 99*(4), 1033-1044.

Chapter 2

Biodiversity and Ecosystem Functions of Sumatran Mangroves Under Anthropogenic Pressures

Mohammad Basyuni[1,2,*]
Shofiyah Sabilah Al Mustaniroh[1]
Faujiah Nurhasanah Ritonga[1,2]
Santy Deasy Siregar[3,4]
and Shigeyuki Baba[5]

[1]Center of Excellence for Mangrove, Universitas Sumatera Utara, Medan, Indonesia
[2]Department of Forestry, Faculty of Forestry, Universitas Sumatera Utara, Medan, Indonesia
[3]Department of Public Health, Faculty of Medicine, Dentistry, and Health Sciences, Universitas Prima Indonesia, Medan, Indonesia
[4]Center of Excellence for PHYTO Degenerative and Lifestyle Medicine, Universitas Prima Indonesia, Medan, Indonesia
[5]International Society for Mangrove Ecosystems, University of the Ryukyus, Nishihara, Okinawa, Japan

Abstract

Mangrove ecosystems in Indonesia are globally significant, covering approximately 2.7 million hectares and representing the largest extent of mangroves worldwide. These ecosystems harbor exceptional biodiversity, serving as habitats for diverse flora and fauna while providing vital ecosystem services such as carbon sequestration, nursery

* Corresponding Author's Email: m.basyuni@usu.ac.id

In: Biodiversity
Editors: Mohammad Basyuni and Faujiah Nurhasanah Ritonga
ISBN: 979-8-90134-246-6

grounds for fisheries, shoreline stabilization, and nutrient cycling. In Sumatra, particularly North Sumatra and South Sumatra Provinces, mangroves play a critical role in sustaining coastal livelihoods and ecological stability. However, rapid land-use change, especially for aquaculture and agriculture, has driven large-scale deforestation and degradation. These pressures, compounded by climate change and unsustainable resource extraction, have reduced forest cover, fragmented habitats, and diminished ecosystem resilience. Biodiversity loss is evident in declining fishery productivity, reduced populations of mangrove-dependent species, and weakened coastal protection against erosion and storms. Despite these threats, restoration and conservation initiatives have demonstrated significant potential. Recent field assessments highlight that replanting programs, community-based rehabilitation, and improved silvicultural practices can accelerate the recovery of ecosystem functions. Mangrove restoration not only enhances biodiversity and carbon storage but also strengthens local livelihoods by supporting sustainable fisheries and ecotourism. Integrating ecological monitoring with traditional knowledge further increases the success of restoration efforts. To ensure long-term resilience, conservation strategies must be embedded within broader coastal management and policy frameworks. Strengthening community participation, enforcing land-use regulations, and promoting sustainable aquaculture practices are essential. Science-driven, adaptive restoration supported by local governance and national policy can secure the ecological integrity of Sumatran mangroves while sustaining their socioeconomic benefits. Protecting and restoring mangroves in Indonesia is not only an ecological necessity but also a strategic investment in climate resilience and sustainable development.

Keywords: Sumatra, biodiversity loss, genetic diversity, aquaculture, deforestation drivers

Introduction

The mangrove forests in Indonesia cover approximately 2.7 million hectares, representing more than 20% of the world's mangrove area (Giri et al., 2011; Basyuni et al., 2022a). These intertidal ecosystems are ecologically and economically valuable, serving as carbon sinks, storm buffers, and hotspots of coastal biodiversity. However, they are under increasing threat from land-use change, pollution, and sea-level rise. Understanding the extent to which these pressures affect biodiversity and ecosystem functioning is critical for sustainable management and policy formulation.

Mangrove ecosystems face significant threats from climate change and anthropogenic pressures, which impact their biodiversity and ecosystem functions across key biogeographic regions. Climate change factors, including sea level rise, increased storminess, and altered precipitation, vary regionally in their impacts on mangrove structure and resilience (Ward et al., 2016). Anthropogenic activities, such as agriculture, aquaculture, and infrastructure development, contribute to mangrove degradation, particularly in densely populated coastal areas such as Southeast Asia (Ng et al., 2025). These pressures can lead to substantial losses in benthic biodiversity, microbial-mediated decomposition rates, and trophic resources (Carugati et al., 2018). Climate change can also disrupt key functional groups, such as the soil microbiota and bioturbators, further eroding biodiversity and ecosystem resilience (Ferreira et al., 2024). To address these challenges, conservation efforts and management strategies must consider region-specific climate change impacts and the responses of key functional groups to ensure the preservation of mangrove ecosystems and their vital ecosystem services.

This chapter aims to comprehensively review the current status, challenges, and impacts of environmental and anthropogenic changes on the biodiversity and ecosystem functioning of Sumatran mangrove forests. With a regional focus on North Sumatra, South Sumatra, and Aceh, this review synthesizes recent ecological assessments and biodiversity surveys to identify key conservation priorities. The objective is to inform future restoration strategies by integrating ecological monitoring, local community participation, and adaptive management approaches to increase the resilience and sustainability of mangrove ecosystems in Sumatra.

Biodiversity of Indonesian Mangroves

Indonesia has the world's highest mangrove species richness, with 240 recorded species, including 48 classified as "true mangroves" and 192 as associated species. The diversity is distributed across multiple plant forms, such as trees, shrubs, herbs, lianas, epiphytes, and palms. In Sumatra, the species count ranges from 70–90, depending on the site and survey scope. The key true mangrove species commonly found include Rhizophora apiculata, Avicennia marina, and Sonneratia alba, which are widespread on Sumatra's east coast.

Table 1. Overview of mangrove biodiversity in Indonesia

Floral Component	Indonesia (National)	Sumatra (Regional)	Reference
Total Mangrove Species	240 species (48 true, 192 associates)	Estimated 70–90 species	Rahman et al., 2024
True Mangrove Species	48 species	Dominant: Rhizophora apiculata, Avicennia marina, Sonneratia alba	Basyuni et al., 2022
Associated Species	192 species	Coastal vines, herbs, and palms in ecotone zones	Rahman et al., 2024
Vegetation Types	74 trees, 36 shrubs, 52 herbs, 6 palms, 43 epiphytes, 23 lianas, 3 ferns, 3 parasites	Similar types observed regionally	Rahman et al., 2024

The floral diversity of Indonesian mangroves reflects the ecological complexity of these ecosystems. Table 1 provides a comparison between national and regional (Sumatra) floristic compositions, emphasizing both true mangroves and associated species. This overview highlights the extraordinary taxonomic richness that underpins the structural and functional roles of mangroves in Sumatra. Nationally, Indonesia hosts approximately 240 mangrove species, including 48 true mangrove species and 192 associated species, making it one of the most floristically rich mangrove nations globally (Rahman et al., 2024; Suhardi et al., 2024). These species are taxonomically diverse and include 74 tree species, 36 shrubs, 52 herbs, 6 palms, 43 epiphytes, 23 lianas, 3 ferns, and 3 parasitic taxa.

In Sumatra, floristic inventories indicate the presence of 70 to 90 mangrove species, which are predominantly composed of true mangroves such as *Rhizophora apiculata*, *Avicennia marina*, and *Sonneratia alba*. These species form the structural backbone of Sumatra's estuarine and coastal mangrove zones and are widely used in local reforestation efforts. Associated species in Sumatra include coastal vines, palms, and understory plants, which increase habitat complexity and support faunal diversity. These data show that while Sumatra's mangrove flora is slightly less diverse than the national total, it includes key foundation species (e.g., Rhizophora apiculata and Avicennia marina) essential for restoration programs. These data show that while Sumatra's mangrove flora is slightly less diverse than the national total, it includes key foundation species (e.g., *Rhizophora apiculata* and *Avicennia marina*) essential for restoration programs.

Despite regional variation, the general composition reflects a strong representation of Indo-Pacific mangrove flora. However, habitat degradation, land-use change, and sedimentation remain key threats to species richness, particularly in western Indonesia.

Table 2 provides a comprehensive summary of faunal diversity in Indonesian mangroves, with a regional focus on North Sumatra. A total of 125 fish species were found nationally, whereas 62 species from 39 families were identified in Langkat and Deli Serdang via morphological analysis. DNA barcoding confirmed that 17 species were affected. Macrozoobenthic diversity included 28 taxa from five phyla. The ecosystems of North Sumatra also support critically endangered species such as Larimichthys crocea and Thunnus maccoyii.

Table 2. Faunal diversity associated with Indonesian mangroves, with an emphasis on North Sumatra

Faunal Group	Species Count (Indonesia)	Notable in North Sumatra (Langkat & Deli Serdang)	Reference
Fish	125 species from 47 families	62 species (39 families, morphology); 17 confirmed via 16S rRNA. Key: Leiognathus equula, Pomadasys argenteus. CR: Larimichthys crocea; EN: Thunnus maccoyii	Basyuni et al. (2025)
Macrozoobenthos	169 species from 52 families	28 taxa from 5 phyla. Dominant: Cerithidea cingulata, Uca spp.	Basyuni et al. (2021)
Birds	80 species	Milky Stork, herons, egrets, kingfishers	Rahman et al., 2024
Reptiles (Squamata)	38 species	Cerberus rhynchops, monitor lizards, geckos	Rahman et al., 2024
Amphibians	6 species	Present in freshwater/brackish transition zones	Rahman et al., 2024
Mammals	21 species	Fruit bats, monkeys, rodents	Rahman et al., 2024
Crocodiles	4 species	Crocodylus porosus	Rahman et al., 2024
Turtles	11 species	Sea turtles nesting near coastal mangroves	Rahman et al., 2024

Faunal assemblages further illustrate the ecological significance of Sumatran mangroves. Table 2 summarizes the major groups, emphasizing North Sumatra as a case study for species richness and conservation status, with detailed data from North Sumatra (Langkat and Deli Serdang). At the national level, Indonesia's mangrove zones support 125 fish species from 47 families, 169 macrozoobenthic species, 80 bird species, and multiple taxa of reptiles, amphibians, mammals, and turtles (Rahman et al., 2024).

In North Sumatra, recent research by Basyuni et al., (2025) recorded 62 fish species from 39 families through morphological identification, with 17 species confirmed by 16S rRNA DNA barcoding, demonstrating the effectiveness of integrating molecular tools with traditional taxonomy. The dominant fish species included Leiognathus equula, Pomadasys argenteus, and Ambassis vachellii, while two IUCN-listed threatened species (Larimichthys crocea – Critically Endangered; Thunnus maccoyii – Endangered) were also identified.

Macrozoobenthic communities include 28 taxa from five phyla (Molluska, Arthropoda, Annelida, Echinodermata, and Sipuncula), with species such as Cerithidea cingulata and Uca spp. dominating restored and natural habitats (Basyuni et al., 2021). This diversity plays a crucial role in sediment stability, nutrient cycling, and food sources for relatively high trophic levels.

The presence of migratory birds, mangrove snakes (Cerberus rhynchops), monkeys, sea turtles, and estuarine crocodiles (Crocodylus porosus) further illustrates the multitrophic ecological value of mangrove systems in North Sumatra. These habitats support both biodiversity and community-based livelihoods, emphasizing their conservation priority.

Table 2 summarizes the faunal diversity associated with Indonesian mangrove ecosystems, with detailed data from North Sumatra (Langkat and Deli Serdang). At the national level, Indonesia's mangrove zones support 125 fish species from 47 families, 169 macrozoobenthic species, 80 bird species, and multiple taxa of reptiles, amphibians, mammals, and turtles (Rahman et al., 2024).

Macrozoobenthic communities include 28 taxa from five phyla (Molluska, Arthropoda, Annelida, Echinodermata, and Sipuncula), with species such as Cerithidea cingulata and Uca spp. dominating restored and natural habitats (Basyuni et al., 2021). This diversity plays a crucial role in sediment stability, nutrient cycling, and food sources for relatively high trophic levels.

The integration of morphological and molecular approaches demonstrates high biodiversity and the presence of threatened species, reinforcing the conservation importance of Sumatran mangroves.

The distribution of mangrove cover varies substantially across Indonesia's islands. Table 3 compares area and species richness, showing Papua as the core mangrove stronghold, whereas Sumatra represents a significant but pressured region. Table 3 also outlines the mangrove area distribution across major Indonesian islands, with a national total of 2,736,985.87 ha. Papua holds the largest share (36%), followed by Kalimantan and Sumatra. Despite their smaller size, Java and Bali-Nusa Tenggara hosts have fragmented yet ecologically vital mangroves.

Table 3. Distribution of mangrove area and species richness across major Indonesian islands (Basyuni et al., 2022a)

Region/Island	Mangrove Area (ha)	% of Total (2,736,985.87 ha)	Species Richness/Key Notes
Papua	984,428	35.96%	Largest mangrove extent; high species richness; intact ecosystems
Kalimantan	857,023	31.32%	Delta-dominated systems; diverse fish and crustacean fauna
Sumatra	416,587	15.22%	Rich in fish and macrozoobenthos; active restoration areas
Sulawesi	200,706	7.33%	Narrow coastal distribution; moderate flora and fauna diversity
Maluku	78,292	2.86%	Relatively undisturbed; includes scattered pristine stands
Bali–Nusa Tenggara	50,749	1.85%	Small fragmented forests; important for local fish nurseries
Java	34,200	1.25%	Most degraded; supports estuarine biodiversity and migratory birds
Indonesia Total	2,736,985.87	100%	48 true mangrove species, 192 associates

The biodiversity of Indonesian mangroves includes 157 species of flora, comprising 52 true mangrove tree species, 21 shrubs, 13 lianas, 7 palms, 14 grasses, 8 herbs, 3 parasites, 36 epiphytes, and 3 ferns. Additionally, there are approximately 122 species of invertebrates, 45 species of fishes, and 148 species of terrestrial fauna. This rich biodiversity supports various ecological functions and livelihoods for millions of coastal communities in Indonesia, despite facing threats from human activities and environmental disturbances (Basyuni et al., 2022a). Sumatra's mangrove area, although smaller than those of Papua and Kalimantan, is highly significant for biodiversity and restoration efforts because of its fragmented but ecologically vital stands.

Key Drivers of Deforestation

Deforestation and habitat degradation in Sumatran mangroves are driven primarily by anthropogenic activities, which significantly impact biodiversity. The main drivers include land conversion for agriculture and aquaculture, illegal logging, and pollution. These activities not only reduce mangrove area but also alter species composition and genetic diversity, leading to a decline in overall ecosystem health. Table 4 shows the key drivers of mangrove deforestation in Sumatra, including land conversion, aquaculture, agriculture, pollution, and climate change. The data highlight how multiple pressures synergistically erode mangrove resilience.

Table 4. Key drivers of deforestation in Sumatran mangroves

Drivers	Impact	Reference
Land-Use Change	Conversion of mangroves to agricultural lands and aquaculture ponds accounts for a significant portion of deforestation, with aquaculture alone responsible for 30% of mangrove loss from 2000 to 2012	Richards & Friess (2016).
	Aquaculture expansion, barren land formation, and oil palm plantations are the primary drivers of mangrove deforestation in North Sumatra from 1990 to 2015.	Basyuni, M., & Sulistiyono, N. (2018, February).
	Conversion of mangroves to agricultural lands and aquaculture is the leading cause of deforestation, accounting for 62% of global losses from 2000 to 2016	Goldberg et al., 2020
	Significant areas of mangroves in Sumatra have been converted into agricultural lands, coconut plantations, and aquaculture ponds, leading to substantial habitat loss and biodiversity decline	Eddy et al., 2021
Agricultural Conversion	Mangroves are cleared for crops such as rice and oil palm, reducing biodiversity and altering hydrological systems.	Goldberg et al. (2020); Giri et al. (2011
Pollution	Industrial and household waste, including plastic debris and heavy metals, contribute to ecosystem degradation	Cahyaningsih et al., 2022
Climate Change	Rising sea levels and altered precipitation patterns disrupt ecological processes, further threatening biodiversity	Ferreira et al., 2024
Oil palm plantation	Expansion of oil palm plantations reduces mangrove cover and fragments habitat, particularly in coastal and estuarine zones.	Basyuni, M., & Sulistiyono, N. (2018).
Barren Land Formation	Deforestation often results in barren, unproductive lands that are unsuitable for restoration, compounding ecosystem loss.	Basyuni, M., & Sulistiyono, N. (2018).
Carbon Emissions	Conversion of mangroves, particularly to coconut plantations, results in high carbon emissions, exacerbating climate change impacts	Eddy et al., 2021

Table 4 presents an overview of the primary drivers of mangrove deforestation in Sumatra, revealing that land-use change is the most dominant factor contributing to forest loss. The conversion of mangrove areas into aquaculture ponds, agricultural fields (particularly rice and oil palm), and coconut plantations has significantly reduced mangrove coverage and altered hydrological systems. Studies have shown that aquaculture alone accounted for up to 30% of mangrove loss in North Sumatra between 1990 and 2015, and globally, it is responsible for more than 60% of deforestation. In addition, industrial and household waste including plastic debris and heavy metals—

has contributed to ecosystem degradation, while climate change effects such as sea-level rise and altered precipitation patterns further disrupt ecological processes. Other contributors include the formation of barren land post deforestation, which often becomes unsuitable for restoration, and the expansion of oil palm plantations that fragment habitats. Notably, the conversion of mangroves into coconut plantations and agricultural land has been associated with increased carbon emissions, undermining the critical role of mangroves in sequestering blue carbon and regulating climate. The dominance of aquaculture and agriculture underscores the urgent need for land-use regulation and integrated coastal management to curb ongoing losses.

The Impact on the Biodiversity of the Sumatran Mangrove

Mangrove deforestation in Sumatra has led to marked declines in biodiversity due to habitat loss, fragmentation, and degradation. Species decline is particularly evident in coastal districts such as Langkat, Deli Serdang, and Serdang Bedagai, where mangrove conversion into aquaculture ponds and settlements has significantly reduced faunal and floral diversity. Several species, including Rhizophora mucronata, Avicennia marina, and mangrove crabs (Scylla spp.), have experienced population pressure, and some are listed as vulnerable or threatened at local scales (Basyuni ct al., 2022a,b).

Furthermore, anthropogenic pressures such as illegal logging, pollution, and fragmentation have resulted in reduced genetic variation among key mangrove species. A study by Basyuni et al., (2012) using microsatellite markers revealed significantly lower genetic diversity in Rhizophora mucronata populations in logged and disturbed mangrove forests in North Sumatra. This decline in genetic variation threatens the adaptive capacity of mangroves to respond to climate change, sea-level rise, and salinity shifts. However, community-based conservation and sustainable use initiatives, such as ecotourism and restoration, have shown promise in restoring biodiversity. In Belawan, North Sumatra, Cahyaningsih et al., (2022) reported that mangrove rehabilitation, coupled with local participation, contributed to increased vegetation cover and species richness in degraded areas.

Table 5. Biodiversity impacts of mangrove deforestation in North Sumatra and documented recovery pathways. The table links species decline, genetic erosion, and habitat fragmentation to successful community-based restoration efforts. The loss of habitat due to deforestation has led to a decline in species richness, including a reduction in mangrove tree species and associated faunal populations such as mud crabs, fish, and macrozoobenthos. In places such as Serdang Bedagai and Deli Serdang, extensive mangrove clearing has resulted in the disappearance of several true mangrove species and a decline in ecological complexity. Fragmentation of mangrove patches along the Langkat and Batubara coasts limits species dispersal and gene flow, whereas molecular studies in North Sumatra show that genetic diversity in Rhizophora mucronata has been significantly reduced in logged areas. Despite these challenges, restoration initiatives that engage local communities, such as those in Belawan, have proven effective in enhancing biodiversity. These community-driven efforts have not only improved vegetation cover but also supported the return of diverse faunal communities, demonstrating the potential for ecological recovery when conservation is integrated with sustainable management practices. This finding reinforces the importance of inclusive and adaptive approaches to mangrove restoration, especially in the face of ongoing environmental and anthropogenic pressures. These findings demonstrate that while deforestation has severe ecological consequences, community-driven restoration projects can restore biodiversity and enhance ecosystem resilience when ecologically appropriate strategies are applied.

Table 5. Biodiversity impacts and recovery potential in North Sumatran mangroves

Impact	Description	Case Example (Location)	Reference
Species Decline/loss	Decrease in mangrove-dependent species due to habitat loss	Serdang Bedagai and Deli Serdang	Ginting et al., 2022
Genetic Variation Loss	Reduced genetic diversity in disturbed mangrove stands, affecting adaptability	*Rhizophora mucronata* in North Sumatra	Basyuni et al., 2012
Habitat Fragmentation	Isolation of mangrove patches limiting gene flow and species dispersal	Langkat and Batubara coasts	Basyuni et al., 2024
Biodiversity Recovery	Community-involved restoration improves species richness and ecological structure	Belawan mangrove rehabilitation	Balke et al., 2021

Conservation and Restoration Efforts

While these conservation strategies are promising, the ongoing pressures from climate change and human activities present significant challenges that require coordinated global efforts to ensure the long-term survival of mangrove ecosystems.

The conservation and restoration of Sumatran mangroves have become increasingly urgent as biodiversity loss and ecosystem degradation intensify. Community-based mangrove restoration has emerged as one of the most effective approaches for reversing deforestation impacts while supporting local livelihoods. In North Sumatra, particularly Belawan, active community involvement in the planting of native mangrove species has led to increased vegetation cover and a resurgence of flora and fauna diversity (Cahyaningsih et al., 2022). These efforts typically prioritize ecologically appropriate species such as Rhizophora apiculata and Avicennia marina, which are known to stabilize shorelines and provide critical nursery grounds for fish and crustaceans (Basyuni et al., 2021). The success of these programs is enhanced when local knowledge, sustainable use practices, and regular monitoring are integrated into the restoration model.

Ecological restoration also plays a vital role in reestablishing ecosystem services, including blue carbon sequestration and flood mitigation. In Lubuk Kertang, North Sumatra, restored mangroves demonstrated significant recovery in biomass and carbon stocksvup to 558 MgC/ha within seven years indicating a strong capacity to rebuild ecosystem function (Amelia et al., 2023). Restoration has also shown positive outcomes for biodiversity, with planted and naturally regenerated mangroves supporting macrozoobenthic diversity and estuarine fish communities essential for ecological resilience (Basyuni et al., 2021; furthermore, restoration, which includes diverse age classes and natural hydrological processes, can better support the recovery of genetic diversity and species dispersal across fragmented landscapes.

Despite these promising developments, ongoing threats such as industrial pollution, poor land-use regulation, and the dominance of monoculture planting continue to hinder broader conservation outcomes. For restoration to be sustainable, it must go beyond tree planting and incorporate ecosystem-based management strategies that account for local socioeconomic contexts, landscape connectivity, and long-term ecological monitoring. Integrating restoration goals into provincial land-use planning and national climate commitments such as Indonesia's nationally determined contributions (NDCs) can align conservation with broader sustainability agendas. Leveraging the ecotourism potential of biodiverse mangrove areas, such as those in Kapo-Kapo Bay, may also provide economic incentives for conservation while promoting awareness of mangrove ecosystem values (Novarino et al., 2023).

Conclusion

Sumatran mangroves are critical ecosystems that sustain high levels of biodiversity and provide key ecological services such as carbon storage, shoreline stabilization, and nursery habitats for marine life. However, deforestation driven by aquaculture expansion, agricultural conversion, pollution, and the impacts of climate change have severely degraded these ecosystems. The consequences include significant species loss, reduced genetic diversity, habitat fragmentation, and the disruption of ecological processes that are vital for both biodiversity and local communities. Despite these challenges, evidence from field studies demonstrates that community-based conservation and restoration initiatives can effectively reverse some of these declines. Replanting native species, engaging local communities, and applying ecologically appropriate silvicultural practices have shown measurable success in enhancing vegetation cover, restoring faunal diversity, and improving ecosystem services such as carbon sequestration and fisheries productivity. These outcomes underscore the importance of integrating traditional knowledge with scientific monitoring to increase the success and resilience of restoration programs. To ensure the long-term sustainability of Sumatran mangroves, restoration and conservation must be embedded in strong policy frameworks and adaptive management strategies. This includes stricter enforcement of land-use regulations, incorporation of mangrove zoning into coastal spatial planning, and promotion of sustainable aquaculture and agriculture. Moreover, future research should prioritize genetic monitoring, hydrological restoration, and socioeconomic assessments to design context-specific interventions. Aligning local restoration with Indonesia's national climate commitments and global blue carbon initiatives will not only safeguard biodiversity but also strengthen climate resilience and support sustainable livelihoods.

Disclaimer

None.

References

Amelia, R., Basyuni, M., Alfinsyahri, A., Sulistiyono, N., Slamet, B., Bimantara, Y., ... & Arifanti, V. B. (2023). Evaluation of plant growth and potential of carbon storage in the restored mangrove of an abandoned pond in Lubuk Kertang, North Sumatra, Indonesia. *Forests*, 14(1), 158.

Balke, T., Vovides, A., Schwarz, C., Chmura, G. L., Ladd, C., & Basyuni, M. (2021). Monitoring tidal hydrology in coastal wetlands with the "Mini Buoy": applications for mangrove restoration. *Hydrology and Earth System Sciences*, 25(3), 1229-1244.

Basyuni, M., Hamzah, H., Rahayu, S., & Siregar, U. J. (2012). Pengaruh Aktivitas Antropogenik Terhadap Keragaman Genetik Rhizophora Mucronata Lamk. Di Hutan Mangrove Secanggang, Sumatera Utara (Effect of Anthropogenic Activities on Genetic Variation of Rhizophora Mucronata Lamk. in Secanggang Mangrove Forest, North. *Foresta*, 1(2), 15365.

Basyuni, M., & Sulistiyono, N. (2018, February). Deforestation and reforestation analysis from land-use changes in North Sumatran Mangroves, 1990-2015. In IOP Conference Series: Materials Science and Engineering (Vol. 309, No. 1, p. 012018). IOP Publishing.

Basyuni, M., Slamet, B., Sulistiyono, N., Munir, E., Vovides, A. G., & Bunting, P. (2021). Physicochemical characteristic, nutrient, and fish production in different types of mangrove forest in North Sumatra and Aceh Provinces of Indonesia. *Kuwait Journal of Science*, 48(3).

Basyuni, M., Sasmito, S. D., Analuddin, K., Ulqodry, T. Z., Saragi-Sasmito, M. F., Eddy, S., & Milantara, N. (2022a). Mangrove biodiversity, conservation and roles for livelihoods in Indonesia. In Mangroves: Biodiversity, livelihoods and conservation (pp. 397-445). Singapore: Springer Nature Singapore.

Basyuni, M., Bimantara, Y., Cuc, N. T., Balke, T., & Vovides, A. G. (2022b). Macrozoobenthic community assemblage as key indicator for mangrove restoration success in North Sumatra and Aceh, Indonesia. *Restoration Ecology*, 30(7), e13614.

Basyuni, M., Aznawi, A. A., Rafli, M., Tinumbunan, J. M. T., Gultom, E. T., Lubis, R. D. A., ... & Baba, S. (2024). Harnessing biomass and blue carbon potential: Estimating carbon stocks in the vital wetlands of Eastern Sumatra, Indonesia. *Land*, 13(11), 1960.

Basyuni, M., Mubaraq, A., Susilowati, A., Rangkuti, A. B., Aznawi, A. A., Al Mustaniroh, S. S., ... & Kajita, T. (2025). Morphological and molecular monitoring of fishes in the waters of mangrove ecosystems. *Global Journal of Environmental Science and Management*, 11(2), 711-732.

Cahyaningsih, A. P., Deanova, A. K., Pristiawati, C. M., Ulumuddin, Y. I., Kusumaningrum, L. I. A., & Setyawan, A. D. (2022). Causes and impacts of anthropogenic activities on mangrove deforestation and degradation in Indonesia. *International Journal of Bonorowo Wetlands*, 12(1).

Carugati, L., Gatto, B., Rastelli, E., Lo Martire, M., Coral, C., Greco, S., & Danovaro, R. (2018). Impact of mangrove forests degradation on biodiversity and ecosystem functioning. *Scientific reports*, 8(1), 13298.

Eddy, S., Milantara, N., Sasmito, S. D., Kajita, T., & Basyuni, M. (2021). Anthropogenic drivers of mangrove loss and associated carbon emissions in South Sumatra, Indonesia. *Forests*, 12(2), 187.

Ferreira, A. C., Ashton, E. C., Ward, R. D., Hendy, I., & Lacerda, L. D. (2024). Mangrove biodiversity and conservation: setting key functional groups and risks of climate-induced functional disruption. *Diversity*, 16(7), 423.

Ginting, Y. R. S., Komarudin, G., & Carr, L. M. (2022). Study of changes in mangrove forest cover in three areas located on the east coast of North Sumatra Province between 1990 and 2020. *Journal of Tropical Forest Science*, 34(4), 467-479.

Giri, C., Ochieng, E., Tieszen, L. L., Zhu, Z., Singh, A., Loveland, T., ... & Duke, N. (2011). Status and distribution of mangrove forests of the world using earth observation satellite data. *Global Ecology and Biogeography*, 20(1), 154-159.

Goldberg, L., Lagomasino, D., Thomas, N., & Fatoyinbo, T. (2020). Global declines in human-driven mangrove loss. *Global change biology*, 26(10), 5844-5855.

Ng, C. Z., Jensen, O. P., & Ong, J. J. (2025). Reconstructing a century of marine landings and fishing effort in Malaysia and Singapore. *ICES Journal of Marine Science*, 82(2), fsaf008.

Nordhaus, I., Toben, M., & Fauziyah, A. (2019). Impact of deforestation on mangrove tree diversity, biomass and community dynamics in the Segara Anakan lagoon, Java, Indonesia: A ten-year perspective. *Estuarine, Coastal and Shelf Science*, 227, 106300.

Novarino, W., Mukhtar, E., Putri, A. S., & Anggaraini, P. L. (2023). Bird diversity and mangrove forest as potential ecotourism destinations in Kapo-kapo Bay, Cubadak Island, West Sumatra, Indonesia. *Biodiversitas Journal of Biological Diversity*, 24(6).

Rahman, Lokollo, F. F., Manuputty, G. D., Hukubun, R. D., Krisye, Maryono, ... & Wardiatno, Y. (2024). A review on the biodiversity and conservation of mangrove ecosystems in Indonesia. *Biodiversity and Conservation*, 33(3), 875-903.

Richards, D. R., & Friess, D. A. (2016). Rates and drivers of mangrove deforestation in Southeast Asia, 2000–2012. *Proceedings of the National Academy of Sciences*, 113(2), 344-349.

Ward, R. D., Friess, D. A., Day, R. H., & Mackenzie, R. A. (2016). Impacts of climate change on mangrove ecosystems: a region by region overview. *Ecosystem Health and sustainability*, 2(4), e01211.

Chapter 3

A Comparison of Sampling Methods for Estimating Mangrove Forest Biodiversity in Selected Indonesian Forest Types: Efficiency and Accuracy Analyses

Andi Aznan Aznawi[1,4]
Mohammad Basyuni[2,3,*]
Luthfi Aziz Mahmud Siregar[4]
and Diana Sofia Hanafiah[4]
[1]Graduate School of Agricultural Science, Faculty of Agriculture, Universitas Sumatera Utara, Medan, Indonesia
[2]Center of Excellence for Mangrove, Universitas Sumatera Utara, Medan, Indonesia
[3]Department of Forestry, Faculty of Forestry, Universitas Sumatera Utara, Medan, Indonesia
[4]Department of Agrotechnology, Faculty of Agriculture, Universitas Sumatera Utara, Medan, Indonesia

Abstract

Mangroves are very important ecosystems for balancing coastal geomorphology. Mangrove forests are home to a wide variety of plant species, animals and other organisms. Each type of mangrove species has different ecological roles, such as carbon sequestration, shoreline protection, and biodiversity provisioning. Vegetation analysis via the transect method is a survey technique used to study vegetation and characterize the environment along a specific straight line. This method

* Corresponding Author's Email: m.basyuni@usu.ac.id

In: Biodiversity
Editors: Mohammad Basyuni and Faujiah Nurhasanah Ritonga
ISBN: 979-8-90134-246-6

involves lines drawn from the seaward zone to the landward zone through various mangrove zones. As a frequently used monitoring method, there are many pitfalls and obstacles that must be considered and avoided, but this method may still be considered effective for use in current field surveys. In this paper, we review the basic methodology and comparison of sampling methods with better analysis and accuracy, systematically covering the application of these methods in global ecology thus far, including determining the starting point of sampling and monitoring the distribution of species in a zonation community. We also discuss appropriate survey methods to be used in different mangrove forest types, ranging from fringing mangrove forests to riverine, basin, and delta areas. The transect method is a universal method used for monitoring species diversity. Compared with random plot sampling, transects were 20% more accurate at detecting species richness but required 30–40% less field effort than grid sampling. However, it needs a development stage along with technological developments and standardization procedures in the field. Therefore, the data generated are more accurate for ecological monitoring and global conservation studies.

Keywords: bioaccessibility assessment, sampling efficiency, mangrove forest

Introduction

Mangroves are highly productive tropical intertidal coastal ecosystems that play an important role in maintaining coastal geomorphological balance. Mangrove forests act as erosion barriers, wave energy dissipators, pollutant filters, and significant blue carbon sinks (Alongi, 2020). Despite occupying less than 1% of the global coastline, mangrove ecosystems contribute approximately 10–15% of the carbon storage in coastal sediments and 10–11% of the terrestrial carbon input into the ocean (Choudhary et al., 2024). In addition to their ecological value, mangrove forests also provide economic and social benefits, such as local livelihoods such as fishing, firewood, honey, and the growing potential for ecotourism (Lee et al., 2014; Ebeler et al., 2025).

The structure and composition of mangrove forests are very important for developing effective conservation strategies. Canopy structure, tree density, diameter at breast height (DBH), and species distribution vary along a gradient from the coastal zone to the mainland (Nugroho et al., 2019). Pioneer species such as *Avicennia marina* and *Sonneratia alba* often dominate the front zone, whereas species such as *Rhizophora apiculata* and *Bruguiera gymnorrhiza* are commonly found in the middle to rear zones (Friess et al., 2012; Njana, 2020).

This information can only be obtained systematically through appropriate field survey methods.

One of the most widely used methods is the line transect method. This method is carried out by drawing a straight line from the sea to the land and dividing the line into observation plots at certain intervals. The transect method allows observations of changes in vegetation structure and composition according to environmental gradients (Vermeersch & Kerckvoorde, 2016). For example, research in the Solomon Islands has used transects with circular plots every 25 meters, recording 12–26 true mangrove species, with a dominance of the Rhizophoraceae family (Bosire et al., 2008; Njana, 2020). Another study in Greece employed a transect-based approach integrating ecological, social, and visual criteria along landscape gradients, enabling participatory mapping and identifying key conservation zones to support climate-resilient landscape planning (Vlami et al., 2023).

Although the transect method is considered a 'universal' sampling method, it has several limitations, such as limited spatial coverage, bias in the selection of starting points, and limitations in capturing the temporal dynamics of mangrove communities. Therefore, the development of transect methods with the integration of technologies such as geographic information systems (GISs) for determining the starting point for sampling is very helpful in obtaining all data on species diversity and analyzing the vegetation structure of mangrove forests (Kuenzer et al., 2011; Hutschenreiter et al., 2021).

Accordingly, this review aims to provide a comparative assessment of transects, random plots, and systematic grid sampling methods with respect to their efficiency and accuracy in estimating mangrove biodiversity. Particular emphasis is placed on the performance of these methods across different Indonesian mangrove forest types, including fringing, riverine, basin, and deltaic systems. This review further identifies the strengths and limitations of the transect approach in capturing vegetation structure, species composition, and ecological zonation while also exploring the integration of modern tools such as GIS and unmanned aerial vehicle (UAV) imagery to increase precision and standardization. Ultimately, this study offers practical recommendations for optimizing sampling strategies to support biodiversity monitoring, conservation planning, and long-term mangrove management.

Forest Structure and Species Composition

The structure of mangrove forests is usually described via several quantitative parameters that reflect the overall condition of the ecosystem. The most commonly used parameters are tree density per hectare, DBH, tree height, and canopy cover percentage. Tree density is a key indicator in assessing vegetation density; mangrove forests with high density tend to have a greater ability to trap sediment and protect coastlines (Sraun et al., 2022).

The diameter at breast height distribution provides an overview of forest regeneration dynamics. A structure dominated by small-diameter trees generally indicates a population that is in an active regeneration phase, whereas the dominance of large trees indicates a mature forest with a lower regeneration rate (Tomlinson, 2016). Tree height and canopy stratification are also indicators of ecosystem health, where mangrove forests with good stratification typically have diverse canopy layers ranging from seedlings to mature trees (Alongi, 2020). Furthermore, the rear zone or transition zone toward the mainland is often occupied by species that are more tolerant to drier soil conditions and lower salt levels, such as *Xylocarpus granatum* or *Lumnitzera racemosa* (Friess et al., 2012; Kusmana et al., 2024). Several studies have also reported the presence of associated species, such as *Nypa fruticans,* in river or estuary zones, indicating unique interactions between hydrological conditions and vegetation composition (Osland et al., 2016).

To synthesize findings across the reviewed studies, a comparative table (Table 1) has been developed. The table highlights differences in efficiency, species detection, and field effort among transect, random, and grid sampling methods, as well as combinations and technological integrations. For example, transect-based surveys in the Solomon Islands detected up to 26 true mangrove species, demonstrating greater sensitivity to zonation patterns than random plot approaches (Bosire et al., 2008). In contrast, random plot sampling, while reducing bias, tends to underestimate species richness by 10–15% relative to transects (Oribhabor & Anyanwu, 2019). Systematic grid sampling provides comparable richness estimates but requires 30–40% more field time (Lovelock et al., 2010). Compared with transect surveys alone, the integration of UAV imagery improved species mapping accuracy and expanded coverage by more than 25% (Kuenzer et al., 2011; Phan et al., 2019; Basyuni et al., 2025). These comparative insights provide a quantitative basis for recommending adaptive sampling strategies across different mangrove types.

Table 1. Comparative efficiency and accuracy of mangrove sampling methods as reported in selected studies

Sampling Method	Typical Plot/Interval Design	Strengths	Limitations	Reported Efficiency/Accuracy	Reference
Transect (line + plots)	20–25 m intervals, 10×10 m or 20×20 m plots; sea–land gradient	Captures zonation, efficient field implementation, clear structural gradients	Bias if transect not representative; limited spatial coverage	Detected up to 26 true mangrove species in Solomon Islands; 20–40% higher sensitivity to zonation patterns vs. random plots	Bosire et al. (2008); Vermeersch & Kerckvoorde (2016)
Random plots	Independent plots (10×10 m) distributed randomly	Reduces location bias, broad representativeness	Less efficient in detecting zonation; may miss rare species	Tends to yield lower spatial resolution, with 10–15% fewer species detected compared to transects	Oribhabor & Anyanwu (2019)
Systematic grid	Grid network (20×20 m or 50×50 m) across site	Broad spatial coverage; allows biomass estimation	Requires high time and labor; difficult in inaccessible mangrove terrain	Comparable species richness to transects but with 30–40% greater field effort	Lovelock et al. (2010)
Transect + Random (combined)	Transects complemented with random plots	Balances zonation and representativeness	Requires more complex design	Provides most comprehensive data on species and biomass distribution	Bosire et al. (2008)
Transect + GIS/UAV integrated	Transects guided by GIS maps, UAV imagery for canopy analysis	High spatial accuracy; repeatable monitoring; reduced field bias	Needs technical capacity and resources	UAV orthophotos improved accuracy of species mapping; enhanced spatial coverage by >25% compared to field-only transects	Kuenzer et al. (2011); Phan et al. (2019); Basyuni et al. (2025)

Several studies have shown variations in this structure across different locations. Research in the Wondama Gulf reported densities of more than 1,000 trees per hectare in the central zone dominated by *Rhizophora apiculata,* with an average DBH of 15–20 cm (Dharmawan & Widyastuti, 2017). In addition, a study on the North coast revealed that rehabilitated mangrove forests have a more homogeneous structure with undeveloped canopy stratification, indicating that stand age and human intervention significantly influence forest structure (Usman et al., 2023). Thus, the structural characteristics of mangrove forests describe not only the current ecosystem conditions but also the history of management and environmental pressures they have experienced. For structural measurements, the best practice recommends the use of 10 × 10 m plots for saplings and seedlings, whereas larger 20 × 20 m plots are more appropriate for mature trees. In basin and deltaic mangroves, where tree density and canopy stratification are relatively high, nested plots (5 × 5 m inside relatively large plots) may be used to capture regeneration dynamics. Standardizing these parameters allows comparisons across sites and improves the robustness of biodiversity assessments (Lovelock et al., 2010; Sasmito et al., 2023).

Variation in Structure and Species Composition Based on Zoning

The species composition of mangrove forests follows a clear zoning pattern from the coastline to the mainland. The front zone or intertidal zone, which is most often submerged by seawater, is usually dominated by pioneer species with high tolerances to salinity and flooding, such as *Avicennia marina*, *Avicennia alba*, and *Sonneratia alba* (Tomlinson, 2016). These species have physiological adaptations, such as salt glands and pneumatophores (breathing roots), that enable them to survive in extreme conditions (Srikant et al., 2015).

In the middle zone, as the flooding frequency decreases and substrate stability increases, the species composition shifts to a dominance of the Rhizophoraceae family, such as *R. apiculata*, *Rhizophora mucronata*, and *Bruguiera gymnorrhiza.* These species are known for their prop roots or knee roots, which help hold sediment in place and strengthen the structure of the forest (Alongi, 2020). This zone generally exhibits moderate to high density with larger trunk sizes than the front zone does. The characteristics of the substrate and the accessibility of freshwater are paramount for the distribution of mangroves. These fluctuations result in distinct growth patterns and zoning configurations, as evidenced in the mangrove ecosystems of Lampung Timur Regency, Indonesia (Imron et al., 2024).

This zoning is influenced not only by environmental factors such as salinity, tides, and sediment supply but also by anthropogenic pressures. A study in the Mekong Delta revealed that large-scale rehabilitation with the planting of *Rhizophora mucronata* resulted in a more homogeneous species composition than natural forests did, thereby disrupting the natural zoning pattern (Phuong et al., 2024).

The Impact of Disturbance and Restoration

Over the past 10 decades, the structure and composition of mangrove forests have been greatly affected by anthropogenic pressures, such as land conversion to fishponds, excessive logging, and coastal development. These disturbances often result in the loss of canopy stratification, a decrease in tree density, and a reduction in species diversity (Alongi, 2020). Additionally, biotic factors such as pests and diseases arise from the degradation of forest health due to rehabilitation efforts that are not prioritized in conservation efforts (Aznawi et al., 2024). In many locations, degraded mangrove forests tend to be dominated by pioneer species with high tolerance to disturbance, such as *Avicennia marina*, which can survive in open soil and high-salinity conditions (Triest et al., 2021).

Restoration programs carried out in various tropical countries have shown varying results. In some locations, restoration involving the planting of a single species, *Rhizophora mucronata*, has resulted in stands with a homogeneous structure, which differs from natural forests, which have diverse species compositions and complex canopy stratification (Tinh et al., 2022). Research in the Mekong Delta shows that although land cover has been successfully restored, species diversity remains low because restoration practices do not consider natural zoning (Nam et al., 2016). The presence of various age strata of trees, from seedlings and saplings to mature trees, is also an indicator of good natural regeneration, which is very important for long-term conservation (Tomlinson, 2016).

Sasmito et al., (2023) reported that the successful restoration of mangroves plays a significant role in achieving the Sustainable Development Goals set by the United Nations. Although mangrove restoration practices have shifted from silvicultural reforestation to hydrological and ecological restoration, the prevailing method of restoration worldwide still involves the planting of monocultures of Rhizophora sp. in Asia and Southeast Asia. Furthermore, monocultures suffer from poor success rates (Friess et al., 2022) and rarely facilitate the restoration of ecosystem services, partly because of their intrinsically lower habitat complexity and limited ability to recover from storm damage (Lee et al., 2019).

In addition, chronic disturbances such as selective logging and domestic waste disposal can alter community structure by increasing the proportion of opportunistic species. A study in India revealed that heavily disturbed areas had lower Shannon diversity indices than protected areas did, indicating that direct human pressure has an impact on forest structure (Rathaude et al., 2025). However, research also shows that when restoration is carried out via an ecosystem-based approach, for example, by considering zoning, hydrology, and species interactions, complex forest structures and diverse species compositions can be restored in the long term (Stanturf et al., 2014). Effective restoration efforts must consider local conditions and natural zoning so that the results not only increase vegetation cover but also approximate the structure and composition of natural mangrove forests (Sasmito et al., 2023).

The Application of Sampling Methods

The transect method is one of the most widely used sampling techniques in mangrove vegetation studies because of its simplicity, efficiency, and ability to capture environmental gradients from sea to land. The basic principle is to draw a straight-line representing changes in environmental conditions along the route and then divide the line into standardized observation plot intervals. Each plot was measured for vegetation parameters such as tree density, diameter at breast height, tree height, and species identification (Bonham, 2013; Friess et al., 2022). On the basis of best practices, plot intervals of 20–25 m are generally sufficient for fringing mangroves with narrow zonation, whereas shorter intervals of 10–15 m are recommended for estuarine and riverine mangroves where species turnover is rapid. The key parameters included species identity, DBH, basal area, canopy cover, and regeneration class (seedlings, saplings, mature trees). To guide method selection, a decision framework can be adopted: (i) use transects with wider intervals in homogenous stands (e.g., rehabilitated areas), (ii) apply nested plots with shorter intervals in heterogeneous natural forests, and (iii) integrate GIS or

UAV imagery to optimize the placement of transects in deltaic systems (Friess et al., 2012; Hutschenreiter et al., 2021).

In practice, the size and distance between plots can be adjusted according to the research objectives. Large plots (e.g., 20 × 20 m) are typically used to observe mature trees, whereas smaller plots (10 × 10 m or 5 × 5 m) are used for growth classes such as saplings and seedlings (Bosire et al., 2008). The transect method allows researchers to monitor species distribution patterns along gradients of salinity, tidal depth, and substrate variation (Alongi, 2020). Additionally, this approach facilitates comparisons between locations because data are collected repeatedly at the same intervals along the transect line.

Another advantage of the transect method is its ability to reveal changes in vegetation communities over relatively short distances. For example, research in the Solomon Islands used line transects with plot intervals of every 25 meters and successfully documented up to 26 true mangrove species, demonstrating the sensitivity of this method to spatial variation (Bosire et al., 2008). On the basis of this fundamental principle, the transect method has been recognized as a universal sampling method that can be applied to various types of mangrove forests, both natural and rehabilitated, provided that standardization is needed to facilitate comparisons of results across studies (Stokes et al., 2023).

Implementation of Transect Methods in Various Forest Types

The transect method can be applied to various types of mangrove forests with adjustments to field conditions, research objectives, and the characteristics of the ecosystem being observed. In coastal mangrove forests, transects are typically drawn from the shoreline toward the mainland to capture the gradient from the pioneer zone, which is often submerged by seawater, to the more stable central zone. Studies in the coastal areas of the Pacitan coast in East Java have shown that this method is effective in identifying the dominance of *Avicennia marina* in the front zone and the transition to *Rhizophora apiculata* in the central zone, with significant changes in tree density along the transect (Sholiqin et al., 2021).

Table 2. Recommended sampling approaches for different mangrove forest types in Indonesia

Mangrove Forest Type	Recommended Sampling Method	Plot/Interval Design	Reported Outcomes	Reference
Fringing mangroves (narrow strips along coast)	Line transect with 20–25 m intervals	10×10 m plots for saplings, 20×20 m for mature trees	Captures clear zonation gradient (Avicennia → Rhizophora); efficient coverage in narrow zones	Sholiqin et al. (2021); Vermeersch & Kerckvoorde (2016)
Riverine mangroves (tidal rivers, freshwater influence)	Transect with shorter intervals (10–15 m) due to rapid species turnover	Nested plots (5×5 m within 10×10 m) for regeneration	Detected shifts from *R. apiculata* to *Sonneratia caseolaris* near estuary; ~15–20% more species recorded with shorter intervals	Stokes et al. (2023)
Basin mangroves (inland depressions with poor flushing)	Systematic grid or combined transect + random plots	Larger 20×20 m plots; additional regeneration subplots	Identified homogeneous stands in rehabilitated sites vs. diverse stratification in natural forests	Usman et al. (2023); Phuong et al. (2024)
Deltaic mangroves (sediment-rich estuaries)	Transect integrated with GIS/UAV imagery	200 m transects with tiered plots; UAV orthophotos for canopy mapping	Found dominance of *R. mucronata* in restored stands; UAV improved spatial coverage by >25%	Phuong et al. (2024); Kuenzer et al. (2011); Basyuni et al. (2025)

The applicability of different sampling methods varies by mangrove forest type (Table 2). In fringing forests, transects with 20–25 m spacing efficiently capture zonation patterns along narrow coastal strips (Sholiqin et al., 2021). Riverine mangroves benefit from shorter intervals (10–15 m), which improve the detection of species turnover driven by freshwater–salinity gradients, increasing species counts by up to 20% compared with wider intervals (Stokes et al., 2023). Basin mangroves, often influenced by rehabilitation, require larger plots and sometimes combined methods (transect + random) to detect differences between homogeneous planted stands and structurally complex natural forests (Usman et al., 2023; Phuong et al., 2024). In deltaic systems, transects integrated with GIS and UAV imagery substantially improve accuracy and coverage, with UAV orthophotos expanding spatial detection by more than 25% relative to field transects alone (Kuenzer et al., 2011; Basyuni et al., 2025).

In estuarine mangrove forests, the implementation of transects requires consideration of tidal dynamics and the influence of freshwater. Research on the Waikaraka estuary revealed that the sharp salinity gradient at the estuary affects species composition, with *R. apiculata* and *Sonneratia caseolaris* being more dominant near the estuary than in other zones (Stokes et al., 2023). Therefore, plot intervals are often shortened (e.g., every 10–15 m) to capture rapid vegetation changes along the gradient.

In mangrove delta forests, transect methods are used to evaluate species distributions in areas with high sediment supplies. Research in the Mekong Delta has applied 200 m transects with tiered plots to monitor the structure of restored forests, revealing a single dominance of *Rhizophora mucronata* in the middle zone and low overall species diversity (Phuong et al., 2024). The implementation of transects in the delta also requires adjustments to the survey timing so that it is conducted during optimal low tide conditions to minimize safety risks and improve accessibility.

In rehabilitated mangrove forests, transects are used to compare regeneration patterns between restored areas and surrounding natural forest areas. A study in Kenya showed that transects in multispecies planted rehabilitation areas resulted in more heterogeneous forest structures and higher diversity indices than transects in areas planted with only one species did (Bosire et al., 2008).

Comparison of the Transect Method with Other Sampling Methods

In practice, this method is often compared with other sampling methods, such as random plot sampling and systematic grid sampling. Each method has advantages and limitations that are important to consider in accordance with the research objectives. Compared with the random plot method, the transect method is more efficient at describing zoning patterns because sampling is carried out linearly following environmental changes. Research by Weisser et al., (2017) revealed that transects can identify changes in species composition more clearly, whereas random plots tend to provide a more homogeneous picture and pay less attention to spatial gradients. On the other hand, random plots are superior in reducing location bias because each point has an equal chance of being selected (Oribhabor and Anyanwu, 2019).

The systematic grid sampling method, in which plots are laid out in a grid pattern across the study area, provides broader spatial coverage than single transects do. However, research in Australia has revealed that grid sampling requires greater resources, both in terms of time and labor, especially in mangrove areas, which are difficult to access (Lovelock et al., 2010). Transects are considered more practical in ecosystems with limited access because they only require linear paths at specific intervals.

Although the transect method has limitations, such as potential bias if the transect line does not represent the heterogeneity of the area, several studies suggest combining methods. For example, Bosire et al., (2008) combined transects with random sampling in specific areas to obtain more comprehensive data on species diversity and biomass distribution. Furthermore, the selection of methods should consider specific objectives, field conditions, and available resources and should not rule out the possibility of integrating several approaches to obtain more robust results (Usman et al., 2023).

Integration of Technology in the Application of Transects

In recent decades, advances in remote sensing technology and GIS have opened new opportunities to improve the effectiveness of transect methods. Traditionally, transects have relied solely on direct field observations, which are time-consuming, labor-intensive, and often limited by site accessibility.

With the integration of technology, transects are not only able to describe the structure and composition of vegetation along observation lines but can also be expanded spatially and temporally.

The use of GIS enables the precise mapping of transect locations and their integration with other spatial data, such as topographic maps, salinity, or sediment distributions. Kuenzer et al., (2011) reported that combining field transects with high-resolution satellite imagery can produce more detailed mangrove zoning maps while identifying critical areas for conservation. Global positioning system (GPS) data linked to GIS also facilitate the replication of sampling in the future, improving data consistency and accuracy (Wang et al., 2019).

Remote sensing through satellite imagery and drone UAVs has been widely used to supplement transect data. Drones can produce orthophotos with very high resolution, enabling the identification of specific species canopies, estimation of land cover areas, and detection of vegetation changes over time (Phan et al., 2019; Larekeng et al., 2024). With these data, field transects can be positioned in representative areas or anomalies detected from the air, thereby increasing survey efficiency (Basyuni et al., 2024).

Field Challenges and Practical Recommendations

Although transect methods are widely used and proven effective for monitoring the structure and composition of mangrove forests, their application in the field often faces various challenges. One of the main obstacles is the accessibility of the location. Many mangrove areas are located in muddy tidal zones that are difficult to access, so setting up transect routes requires careful planning and tools such as boats or footboards. Tidal conditions are also an important factor; surveys conducted during high tide can reduce visibility and increase safety risks for the survey team (Lovelock et al., 2010).

Unpredictable weather conditions, such as heavy rain or extreme temperatures, can affect survey timing and the quality of the data obtained. Research in Australian tropical mangrove forests reported that extreme weather conditions caused survey delays of up to several days and increased bias due to differences in environmental conditions between sampling times (Alongi, 2020). In addition, substrate variability (deep mud, dense prop roots) can make it difficult to set up plots or measure DBH consistently. Another challenge is standardizing methods. Plot size, plot spacing, and measured parameters often vary among studies, making it difficult to compare results across locations or compile comprehensive meta-analyses (Bosire et al., 2008). This highlights the importance of more detailed and standardized technical guidelines, especially for large-scale conservation projects.

To overcome these challenges, several practical recommendations can be implemented. First, survey timing should be adjusted to the tidal schedule and rainy season and supplemented with regular weather monitoring. Second, field team training needs to be improved, including consistent measurement techniques and the use of GPS equipment or GIS-based field applications to minimize recording errors (Wang et al., 2019). Third, the use of adaptive transect methods, such as adjusting plot sizes in hard-to-reach areas or utilizing drones to supplement field data, is recommended.

A deep understanding of the structure and composition of mangrove forests through the transect method has yielded many important findings, but there are still several research gaps that can be addressed. The main gap lies in the lack of consistent standardization in the application of transect methods. The plot size, distance between plots, and measured parameters vary across studies, making comparisons across locations and large-scale meta-analyses difficult. In addition, the integration of modern technologies such as GISs and drones is rarely applied in transect surveys, even though these technologies have been proven to improve data collection accuracy and efficiency. Studies in rehabilitation areas are also limited; most research has focused on natural forests, whereas transect data from restored forests often show species homogeneity due to single-species planting practices that do not consider natural zoning. On the other hand, practical guidelines that are adaptive to local conditions, including field access, tidal factors, and weather, are still minimal, making the application of transects in some areas a challenge in itself.

A number of important findings have contributed significantly to this transect method study. The transect method has proven effective in mapping the ecological gradient from the front to the back of mangrove forests, characterized by shifts in species dominance, such as *Avicennia* in the front zone, *Rhizophora* in the middle zone, and *Xylocarpus* in the back zone. Variations in structure and composition have also been identified in various locations, such as Australia, Vietnam, Indonesia, and the Solomon Islands, which are influenced by differences in geomorphology and anthropogenic pressures. Research shows that human disturbance and restoration programs can alter forest structure, resulting in species homogeneity and simple canopy stratification compared with more complex natural forests. The integration of cutting-edge technology into transect surveys is beginning to show great potential in improving spatial coverage and data accuracy. Additionally, various practical recommendations, such as selecting the appropriate survey timing, training field teams, and adjusting plot sizes, have been proposed to address field challenges.

Challenges and Limitations

The application of the transect method to mangrove ecosystems often faces various technical challenges that impact survey effectiveness. One major challenge is site accessibility. Many mangrove areas are located in tidal areas with deep, root-dense muddy substrates, making it difficult for survey teams to move around and establish transect plots. These conditions often require the use of small boats, stepping boards, or other aids to reach observation points, which in turn increases survey time and costs. Furthermore, tidal variability impacts field conditions; data collection during high tide can hinder stem diameter measurements and species identification because some vegetation is submerged. Extreme weather, such as heavy rain or intense heat, can also impact data consistency by reducing sampling time and increasing safety risks for field teams. Other technical challenges include substrate heterogeneity, which can make it difficult to determine precise plot boundaries, and disturbances from local fauna such as crabs or sea snakes, which are often found in mangrove areas. This combination of challenges requires careful field planning, including adjusting plot sizes, selecting appropriate survey times, and implementing strict safety protocols to ensure optimal application of the transect method in the field.

Another major challenge in the application of transects is the difficulty of capturing temporal dynamics in mangrove ecosystems. Because transects often provide only a "snapshot" of community structure, they may fail to reflect regeneration patterns, successional processes, or the impacts of climate variability unless repeated at regular intervals. Long-term monitoring through repeated transect surveys can provide valuable insights into forest recovery following disturbances, including natural regeneration after cyclones or human-induced degradation (Alongi, 2020; Triest et al., 2021). To complement field-based replication, integrating transects with remote sensing time series such as Landsat or Sentinel imagery offers a cost-effective means of tracking canopy cover, species zonation shifts, and biomass changes over time (Kuenzer et al., 2011; Wang et al., 2019). UAV-based monitoring has also been shown to enhance spatial detail, allowing the detection of subtle vegetation changes between survey years (Phan et al., 2019; Larekeng et al., 2024). These approaches address the limitations of static data collection by enabling temporal comparisons and strengthening the predictive value of transect-based studies.

In addition to technical challenges in the field, the transect method has several methodological limitations that must be considered when interpreting research results. First, this method relies heavily on the representativeness of the selected transect lines. If transect lines are not drawn across the full environmental gradient or through areas of high species diversity, the resulting data can be biased and not reflect the overall condition of the ecosystem. Second, varying plot sizes and intervals between studies make cross-site comparisons difficult, as differences in sampling scale can result in differences in density, frequency, and diversity indices that do not always reflect true ecological differences. Third, the transect method tends to be static; the data collected only provide a snapshot of conditions at a specific point in time, making long-term dynamics such as natural regeneration, vegetation succession, or the impacts of climate change impossible to capture directly without periodic replication. Fourth, this method generally focuses on vegetation parameters such as stem diameter, density, and species identification but rarely incorporates other environmental variables such as water quality, nutrient content, or geomorphological factors at depth, even though these variables significantly influence mangrove community structure. Thus, although the transect method is effective for baseline surveys, these methodological limitations need to be considered and complemented with other approaches or additional data integration to produce a more comprehensive analysis.

A further methodological limitation of the transect approach is its static nature, which restricts the ability to monitor changes across time. This limitation can be addressed by designing repeated transect surveys at consistent seasonal or annual intervals, allowing researchers to document regeneration rates, succession trajectories, and responses to climatic events such as sea-level rise or storm surges (Alongi, 2020; Friess et al., 2022). The integration of repeated ground-based transects with time series satellite imagery (e.g., Landsat or Sentinel) and UAV monitoring provides a scalable framework for long-term assessment (Kuenzer et al., 2011; Wang et al., 2019; Larekeng et al., 2024). Such integration enhances comparability between sites and reduces uncertainty in distinguishing natural variability from anthropogenic impacts. Incorporating this temporal dimension is critical for evaluating the resilience and stability of mangrove ecosystems under both restoration and conservation scenarios.

While monitoring in restoration areas has traditionally focused on structural attributes such as species composition, DBH, and canopy stratification, there is growing recognition of the need to integrate functional and ecosystem service indicators into transect-based surveys. For example, measurements of above- and belowground carbon stocks provide insight into the contribution of restored stands to blue carbon sequestration (Sasmito et al., 2023; Choudhary et al., 2024). Indicators of sediment accretion and erosion control can be assessed through transect-linked geomorphological surveys, which reflect the capacity of rehabilitated forests to stabilize coastlines (Lee et al., 2019). Additionally, monitoring the return of associated fauna such as nekton, crab, and avifauna offers a proxy for habitat functionality and biodiversity services (Bosire et al., 2008; Osland et al., 2016). The incorporation of such functional measures alongside structural assessments ensures that restoration success is evaluated not only in terms of vegetation cover but also in terms of its capacity to restore the ecological processes and ecosystem services critical for local communities.

Future Directions

The development of mangrove ecological studies has shown that the transect method remains relevant as a primary tool for monitoring vegetation structure and composition. However, to address global conservation challenges and the need for more precise data, this method needs further development. First, efforts are needed to standardize more detailed and internationally recognized transect protocols, including plot sizes, intervals, and measured parameters. With standards, research results from different locations can be directly compared and used for meta-analyses or large-scale modeling. Second, the integration of digital technology is an important research direction. The application of GIS and remote sensing, such as high-resolution satellite imagery and drones (UAVs), can expand the scope of transect data from local points to broader landscapes. Field data generated from transects can be calibrated with aerial imagery to obtain more detailed and accurate species zoning maps. This integration also allows for regular monitoring of vegetation changes without the need for intensive field surveys every time. Third, a multidisciplinary approach needs to be implemented by combining vegetation transect data with other environmental variables, such as water quality, sediment characteristics, and socioeconomic factors of the surrounding community. With this approach, the analysis results not only describe vegetation conditions but also provide insights into the causes of change and more targeted management recommendations. Fourth, future research is expected to focus more on rehabilitation areas. Transects at multispecies restoration sites, for example, can provide a basis for evaluating the long-term success of restoration compared with that of natural forests. This will help formulate planting policies that not only maintain land cover but also maintain ecosystem quality closer to its original state.

Future studies should place greater emphasis on integrating temporal monitoring and functional indicators into mangrove sampling designs. Establishing long-term monitoring programs that combine repeated transect surveys with satellite or UAV-based time series imagery will allow researchers to capture dynamic processes such as regeneration, successional change, and climate-driven impacts (Wang et al., 2019; Larekeng et al., 2024). Equally important is the incorporation of ecosystem service indicators, including carbon sequestration potential, sediment stabilization, and habitat provision for fauna (Choudhary et al., 2024; Sasmito et al., 2023). These additions move beyond structural assessments to reflect the ecological functionality of mangrove ecosystems. Standardizing such multidimensional approaches across sites will not only enhance scientific comparability but also provide policymakers and practitioners with more comprehensive evidence to guide conservation and restoration strategies at both the national scale and the global scale. Overall, the development of transect methods will shift from conventional approaches to more innovative, standardized, and technology-based approaches. With this approach, transect data will not only be useful for academic research but also become a crucial instrument for supporting mangrove conservation and management at the local and global levels.

Conclusion

Through the transect method, various studies have successfully mapped vegetation zonation patterns, identified dominant species in each zone, and described variations in forest structure across locations. The transect method remains relevant as a universal tool in mangrove vegetation studies, but with a combination of spatial planning and good field planning, its collection and contribution to global conservation efforts can be maximized. Moreover, there is an urgent need for the development of internationally standardized transect protocols, including plot sizes, interval distances, and measured parameters, to ensure the comparability of results across different geographic regions and long-term monitoring initiatives. The integration of modern technologies such as GIS, high-resolution satellite imagery, and UAVs can substantially improve the accuracy, efficiency, and scalability of transect surveys, thereby enabling their application to large and complex mangrove landscapes. Equally important, transect-based studies should be complemented by a multidisciplinary approach that integrates ecological data with abiotic factors such as hydrology, soil, and salinity, as well as socioeconomic factors, ensuring that mangrove conservation and restoration strategies are grounded in both ecological integrity and community needs.

References

Alongi, D. M. (2020). Global significance of mangrove blue carbon in climate change mitigation. *Science*, 367(6478): 129–130.

Aznawi, A. A., Basyuni, M., Hanafiah, D. S., Larekeng, S. H. & Mulya Sari, S. P. (2024). Impact of Pest and Diseases on Mangrove Forest Rehabilitation in Indonesia: A Review. *OnLine Journal of Biological Sciences*. 24(4): 728-738.

Basyuni, M., Amelia, R., Aznawi, A. A., Wirasatriya, A., Iryanthony, S. B., Slamet, B., Musatniroh, S., Rahmania, R., Rahmila, Y. I., Sumarga, E., Larekeng, S. H., Salmo III, S., Kajita, I., Sivaipram, I., Ali, H. A. (2025). Reduction of mangrove carbon stocks ecosystems attributed to illegal logging using combination of unmanned aerial vehicle imagery and field surveys. *Global Journal of Environmental Science and Manage*ment, 11(1): 225-242.

Bonham, C. D. (2013). Measurement for terrestrial vegetation. *Second Edition.* Colorado State University, Fort Collins, Colorado.

Bosire, J. O., Dahdouh Guebas, F., Walton, M., Crona, B. I., Lewis III, R. R., Field, C., Kairo, J. G., & Koedam, N. (2008). Functionality of restored mangroves: A review. *Aquatic Botany*. 89(2): 251–259.

Choudhary, B., Dhar, V., & Pawase, A. S. (2024). Blue carbon and the role of mangroves in carbon sequestration: Its mechanisms, estimation, human impacts and conservation strategies for economic incentives. *Journal of Sea Research*. 199: 102504.

Dharmawan, I. W. E., & Widyastuti, A. (2017). Pristine mangrove community in Wondama Gulf, West Papua, Indonesia. *Marine Research in Indonesia*. 42(2): 1-10.

Ebeler, L., Balke, T., Laurie, E., Le, H., Nguyen, T., & Renaud, F. G. (2025). Local-scale impacts of mangrove restoration and conservation on coastal communities in two Vietnamese deltas: Socioeconomic and institutional dynamics. *Ocean & Coastal Management*. 269: 107812.

Friess, D. A., Krauss, K. W., Horstman, E. M., Balke, T., Bouma, T. J., Galli, D., & Webb, E. L. (2012). Are all intertidal wetlands naturally created equal? Bottlenecks, thresholds and knowledge gaps to mangrove and saltmarsh ecosystems. *Marine Pollution Bulletin*. 64(7): 1339–1352.

Friess, D. A., Adame, M. F., Adams, J. B., & Lovelock, C. E. (2022). Mangrove forest under climate change in a 2°C world. *WIREs climate change*. 13(4): e792.

Hutschenreiter, A., Ramos-Fernandez, G., & Aureli, F. (2021). Line-transect versus point-transect sampling: the effects of survey area and survey effort on method efficiency for Geoffroy's spider monkeys. *Wildlife Research*. 48(7): 590-597.

Imron, Duryat, Irawan, S. E., Maryono, T., Rodiani. (2024). Zoning patterns of mangrove forests in Lampung Timur Regency. *Journal of Sylva Indonesiana*, 7(1), 40–47.

Kuenzer, C., Bluemel, A., Gebhardt, S., Quoc, T. V., & Dech, S. (2011). Remote sensing of mangrove ecosystems: A review. *Remote Sensing*. 3(5): 878–928.

Kusmana, C., Riyadi, R., & Suheri, A. (2024). Mangrove forest restoration in the western part of Rambut Island wildlife reserve, Seribu Islands. *Journal of Natural Resources and Environmental Management*. 14(4): 770.

Larekeng, S. H. Nursaputra, M., Mappiasse, M. F., Ishak, S., Basyuni, M., Sumarga, E., Arifanti, V. B. Aznawi, A. A., Rahmila, Y. I., Yulianti, M., Rahmania, R., Mubaraq, A., Salmo III, S. G., Ali, H. M., Yeni I. (2024). Estimation of mangrove carbon stocks using unmanned aerial vehicle over coastal vegetation. *Global J. Environ. Sci. Manage.*, 10(3): 1133-1150.

Lee, S. Y., Primavera, J. H., Dahdouh Guebas, F., McKee, K., Bosire, J. O., Cannicci, S., Diele, K., Fromard, F., Koedam, N., Marchand, C., Mendelssohn, I., Mukherjee, N., & Record, S. (2014). Ecological role and services of tropical mangrove ecosystems: A reassessment. *Global Ecology and Biogeography*. 23(7): 726–743.

Lee, S. Y., Hamilton, S., Barbier, E. B., Primavera, J., & Lewis, R. R. (2019). Better restoration policies are needed to conserve mangrove ecosystems. *Nature Ecology and Evolution*. 3: 870-872.

Lovelock, C. E., Feller, I. C., Reef, R., Hickey, S., & Ball, M. C. (2010). Mangrove dieback during fluctuating sea levels. *Scientific Reports*. 1–7.

Nam, V. N., Sasmito, S. D., Murdiyarso, D., & Purbopuspito, J. (2016). Allometric equations for aboveground and belowground biomass estimation in mangrove forests of the Mekong River Delta. *Forest Ecology and Management*. 368: 137–146.

Njana, M. A. (2020). Structure, growth, and sustainability of mangrove forests of mainland Tanzania. *Global Ecologi and Conservation*. 24: e01394.

Nugroho, T. S., Fahrudin, A., Yulianda, F., & Bengen, D. G. (2019). Structure and composition of riverine and fringe mangroves at Muara Kubu protected areas. *AACL Bioflux*. 12(1):378-393.

Oribhabor, C. B., & Anyanwu, C. A. (2019). Research sampling and sample size determination: A practical application. *Journal of Education Research*. 2(1):47-56.

Osland, M. J., Feher, L. C., Griffith, K. T., Cavanaugh, K. C., Enwright, N. M., Day, R. H., Stagg, C. L., Krauss, K. W., Howard, R. J., Grace, J. B., Rogers, K., & Kennedy, J. P. (2016). Climatic controls on the global distribution, abundance, and species richness of mangrove forests. *Ecological Monographs*. 87(2): 341–359.

Phan, T. N., Kuch, V., & Lehnert, L. W. (2019). Land cover classification using Google Earth Engine and random forest classifier—The role of image composition. *Remote Sensing*. 11(5): 603.

Phuong, T. T., Vien, T. D., Son, C. T., Thuy, D. T., & Greiving, S. (2024). Impact of climate change on agricultural production and food security: A case study in the Mekong River Delta of Vietnam. *Sustainability*. 16(17): 7776.

Rathaude, S. K., Irtiqa, S., Rahim, P. A., Wani, Z. A., Kothandaraman, S., & Dar, J. A. (2025). Disturbance intensity drives structural, compositional and diversity attributes in tropical dry forests of Central India. *Trees*, Forest and People. 20: 100864.

Sasmito, S. D., Basyuni, M., Kridalaksana, A., Sasmito, M. F. S., Lovelock, C. E., & Murdiyarso, D. (2023). Challenges and opportunities for achieving Sustainable Development Goals through restoration of Indonesia's mangroves. *Nature Ecology and Evolution*. 7(1): 62–70.

Sholiqin, M., Pramadaningtyas, P. S., Solikah, I., Febriyanti, S., Pambudi, M. D., Mahartika, S. B., Umam, A. F., Liza, N., & Setyawan, A. D. (2021). Analysis of the diversity and evenness of mangrove ecosystems in the Pacitan Coast, East Java, Indonesia. *International Journal Bonorowo Wetlands*. 11(2): 84-94.

Sraun, M., Bawole, R., Marwam J., Sinery, A. S., & Cabuy, R. L. (2022). Diversity, composition, structure and canopy cover of mangrove trees in six locations along Bintuni riverbank, Bintuni Bay, West Papua, Indonesia. *Biodiversitas*. 23(11): 5835-5843.

Srikant, S., Lum, S., & Chen, Z. (2015). Mangrove root: adaptations and ecological importance. *Trees*. 30(2).

Stanturf, J. A., Palik, B., & Dumroese, R. K. (2014). Contemporary forest restoration: A review emphasizing function. *Forest Ecology and Management*. 331(1): 292-323.

Stokes, D. J., Glover, H. E., Bryan, K. R., & Pilditch, C. A. (2023). Environmental state of a small intertidal estuary a decade after mangrove clearance, Waikaraka Estuary, Aotearoa New Zealand. *Ocean & Coastal Management*. 243(1): 106731.

Tomlinson, P. B. (2016). *The botany of mangroves* (2nd ed.). Cambridge University Press.

Tinh, P. H., MacKenzie, R. A., Hung, T. D., Vinh, T. V., Ha, H. T., Lam, M. H., Hanh, N. T. H., Tung, N. X., Hai, P. M., & Huyen, B. T. (2022). Mangrove restoration in Vietnamese Mekong Delta during 2015-2020: Achievements and challenges. *Marine Conservation and Sustainability*. 9.

Triest, L., Socorro, A. D., Gado, V. J., Mazo, A., & Sierens, T. (2021). *Avicennia* genetic diversity and fine-scaled structure influenced by coastal proximity of mangrove fragments. *Frontiers Marine*. 8.

Usman, U., Hilmi, E., & Iqbal, A. (2023). The distribution of mangrove area, mangrove density, and species diversity on the North Coast of Jakarta. *Omni-Akuatika*. 19(1): 88.

Vermeersch, S., & Kerckvoorde, A. V. (2016). The use of line transects for an accurate description of river bank vegetation. *Wetlands Ecology and Management*. 24(6).

Vlami, V., Kokkoris, I. P., Charalampopoulus, I., Doxiadis, T., Giannakopoulos, C., & Lazoglou, M. (2023). A transect method for promoting landscape conservation in the climate change context: A case-study in Greece. *Sustainability*. 15(17): 13266.

Wang, L., Jia, M., Yin, D., & Tian, J. (2019). A review of remote sensing for mangrove forests: 1956–2018. *Remote Sensing of Environment*. 231: 111223.

Weisser, W. W., Roscher, C., Meyer, S. T., Ebeling, A., Luo, G., Allan, E., Beßler, H., Barnard, R. L., Buchmann, N., Buscot, F., Engels, C., Fischer, C., Fischer, M.,

Gessler, A., Gleixner, G., Halle, S., Hildebrandt, A., Hillebrand, H., de Kroon, H., Lange, M., ... Eisenhauer, N. (2017). Biodiversity effects on ecosystem functioning in a 15-year grassland experiment: Patterns, mechanisms, and open questions. *Basic and Applied Ecology*. 23. 1–73.

Chapter 4

The Microbial Regulation of Greenhouse Gas Fluxes in Mangrove Sediments: Bibliometric Analysis and Mechanisms

Alfian Mubaraq[1,2]
Mohammad Basyuni[2,3,*]
Tengku Sabrina[4]
and Mariani Sembiring[4]
[1]Graduate School of Agricultural Science, Faculty of Agriculture, Universitas Sumatera Utara, Medan, Indonesia
[2]Center of Excellence for Mangrove, Universitas Sumatera Utara, Medan, Indonesia
[3]Department of Forestry, Faculty of Forestry, Universitas Sumatera Utara, Medan, Indonesia
[4]Department of Agrotechnology, Faculty of Agriculture, Universitas Sumatera Utara, Medan, Indonesia

Abstract

Mangrove habitats are special ecosystems with the capacity to store carbon and reduce greenhouse gas emissions since they are found where land and ocean meet. Furthermore, because of their geographic position, mangrove ecosystems feature intricate biogeochemical processes that support a range of ecosystem services, including carbon sequestration, pollutant containment, and habitats for other organisms. Biogeochemical mechanisms that occur in mangrove sediments include several carbon, nitrogen and sulfur cycles that play a role in the production and consumption of greenhouse gases such as methane (CH_4), carbon dioxide

* Corresponding Author's Email: m.basyuni@usu.ac.id

In: Biodiversity
Editors: Mohammad Basyuni and Faujiah Nurhasanah Ritonga
ISBN: 979-8-90134-246-6

(CO_2), and nitrous oxide (N_2O). This cycle is strongly influenced by the role of microbial communities in changing biogeochemical elements in sediments. A total of 329 documents were bibliometrically evaluated to examine publishing trends, patterns, and correlations among relevant studies on bacteria that contribute to greenhouse gas emissions in mangrove ecosystems. The nitrification mechanism is the most extensively studied aspect of mangrove ecosystems, with implications for mangrove ecosystem management. In addition, attributions, affiliations, source journals and citation analyses were also performed for scientific literature mapping and future research directions. Deciphering the microbial process of its function in the greenhouse gas cycle, as well as the factors that control it, will probably reveal mangrove ecosystem management and the issues it would confront in mitigating climate change mediated by microorganisms.

Keywords: bibliometric, denitrification, microbial regulation, mangrove sediments, greenhouse gases

Introduction

Greenhouse gases are among the most pressing environmental challenges facing the twenty-first century. The greenhouse effect occurs when solar heat is reflected off the Earth's surface, and greenhouse gases absorb and trap heat, ultimately affecting the global climate. Carbon dioxide (CO_2), methane (CH_4), and nitrous oxide (N_2O) are the primary greenhouse gases driving global climate change (Shiau and Chiau, 2020). Since 1750, greenhouse gas concentrations have increased by 144% for CO_2, 256% for CH_4, and 121% for N_2O, according to the World Meteorological Organization (Das and Mandal, 2022).

The mangrove habitat is essential for controlling greenhouse gas emissions. Mangrove ecosystems have the capacity to sequester carbon and contribute to the absorption and storage of greenhouse gases. On the other hand, mangrove sediments, which are rich in organic matter, play a role in promoting and regulating greenhouse gas emissions, whose biogeochemical cycles are driven by microorganisms (Rosentreter et al., 2018). In mangrove ecosystems, bacteria play crucial roles in various biogeochemical cycles, such as those involving carbon, nitrogen, sulfur, and phosphorus (Thatoi et al., 2013; Sengupta and Mitra, 2020).

In mangrove ecosystems, microorganisms carry out biogeochemical cycles, particularly the nitrogen cycle, which includes nitrification, denitrification, nitrogen fixation, mineralization, anammox, comammox, and dissimilatory nitrate reduction to ammonium (DNRA) (Ma et al., 2024; Mubaraq et al., 2024; Qian et al., 2024). Methanogens and methanotrophs, which are involved in the carbon cycle, are also found in mangrove sediments (Lin et al., 2025). Methanogenesis is a metabolic process carried out by bacteria to produce methane, one of the most potent greenhouse gases, which uses hydrogen, carbon dioxide, and methanol under anoxic conditions (Mayer et al., 2019). Methane oxidation can be carried out by methanotrophic bacteria in mangrove ecosystems and has been shown to successfully oxidize more than 50% of the methane produced. The reclamation of mangrove ecosystems for fishponds results in a reduction in methane oxidation by methanotrophs and increases methane emissions into the atmosphere (Yang et al., 2025).

Mangrove sediment conditions, greenhouse gas emissions, and the ecological balance are influenced by complex interactions between the bacteria involved (Holguin et al., 2006). The mechanisms of microorganisms in mangrove sediments are important for understanding the biogeochemistry that influences the greenhouse gas effect (Ghizelini et al., 2012). Thus, this work was carried out as an initial effort to employ bibliometric analysis and review the mechanisms of greenhouse gas driving and regulation carried out by microbes in mangrove sediments, which are significant ecosystems and serve as habitats for a variety of microorganisms. The performance of publications and research trends on bacteria that contribute to greenhouse gas emissions in mangrove sediments was evaluated via bibliometric analysis (Ellegaard and Wallin, 2015). This analysis can be useful in designing initial research on microbial regulation in regulating greenhouse gas emissions in mangrove sediments and recognizing the importance of mangrove ecosystems in climate change.

Methods

Bibliometric analysis was conducted in this study on the specified literature. Bibliometric analysis is a method of analyzing large amounts of published data (Donthu et al., 2021). In July 2025, publication data were gathered from journal databases via the Scopus database. Several keywords were used in the searches to find all or one of the entered keywords. The keywords used were ("Methanotroph") OR ("Methanogens") OR ("Denitrification") OR

("Nitrification") OR ("DNRA") OR ("Anammox") OR ('Comammox') AND ("Mangrove").

Data extraction was performed by first entering the keywords for the search TITLE-ABS-KEY AND (LIMIT-TO (DOCTYPE, "article") OR LIMIT-TO (DOCTYPE, 'review') ORLIMIT-TO (DOCTYPE, "conference paper") OR LIMIT -TO (DOCTYPE, "chapter") OR LIMIT-TO (DOCTYPE, 'book')) AND (LIMIT-TO (LANGUAGE, "English")). Restrictions on document type and language were applied to obtain more accurate results. Data extraction was then performed by exporting the data to CSV format for bibliometric analysis.

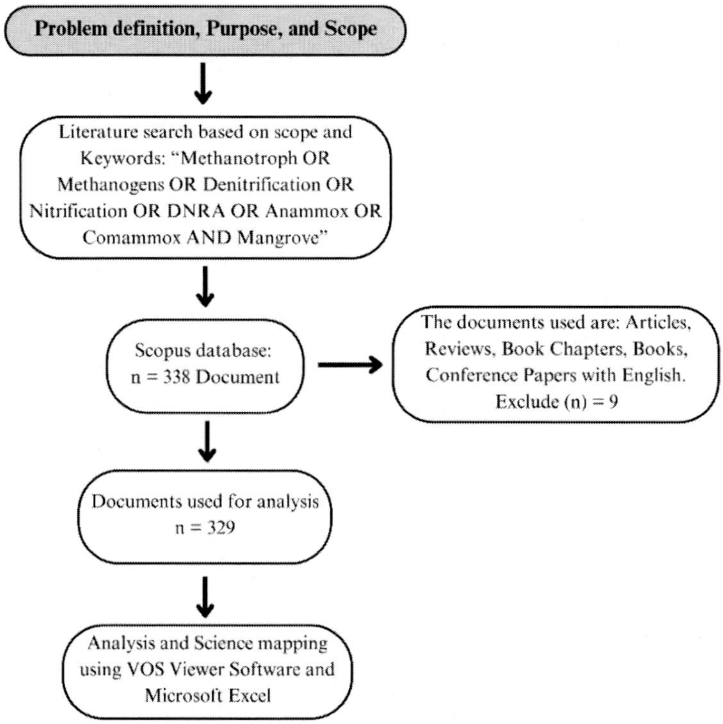

Figure 1. Research framework using bibliometric analysis.

From the extracted data, we used 338 documents from the restrictions and filtered them by excluding 9 documents, resulting in 329 Scopus documents that were analyzed bibliometrically. Key sections of each document were successfully extracted, including i) authors, ii) title, iii) year, iv) source title, v) citation, vi) affiliation, vii) abstract, viii) author keywords, ix) document type, and other supporting data. In this study, a bibliometric analysis was conducted to evaluate research trends, topics, and author affiliations (Liao et al., 2018). VOS Viewer software was used in the mapping analysis and to visualize the data (Van Eck and Waltman, 2010).

Results

Trend Analysis of Scientific Publications

The extracted data were analyzed for publication trends that discuss microorganisms that play a role in greenhouse gas emissions in mangrove ecosystems. Relevant publications were first published in 1982, with a total of one publication, and there were fluctuations in publications until 2010. Until 2011, publications increased significantly, with a total of 12 publications until 2024, with a total of 40 publications. This trend highlights the increasing number of scientific publications in this field and highlights the importance of this field and its rapid progress.

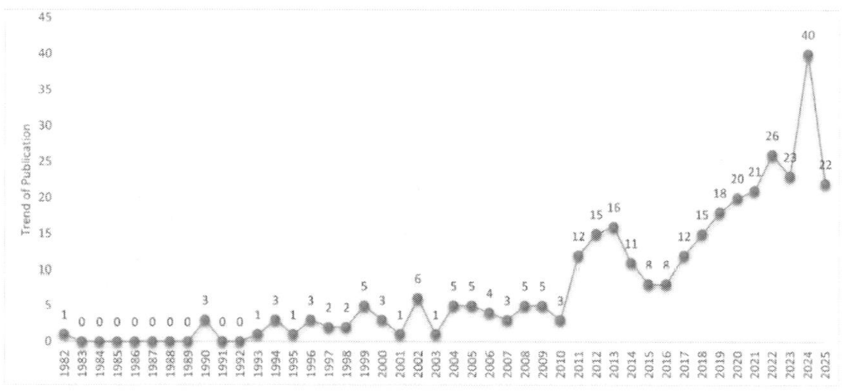

Figure 2. Trend publication of microbes involved in greenhouse gas emissions in mangrove sediments.

Annual Publications and Contributing Countries

The visualization is performed on the coauthorship relationship of each country and shows publication productivity (Figure 2). The size of the nodes indicates the number of publications from each country. Different node colors indicate the country clusters formed on the basis of the publication co-authorship relationship. China, as the country with the highest number of publications, has the same cluster as the Netherlands and Colombia. Indonesia has a different cluster than China does, and Indonesia is in one cluster with Thailand, South Africa, Sri Lanka, Italy, Australia, and Sweden.

Table 1 presents data listing 20 countries with high productivity in publications related to microbial greenhouse gases in mangrove sediments. China contributed the greatest number of publications, with a total of 126 documents out of a total of 329 documents, followed by the United States, with a total of 58, and Indonesia, with 17 documents. China has the highest number of citations, with a total of 3659, followed by the United States, with 2879 citations, and Australia, with 2816 citations.

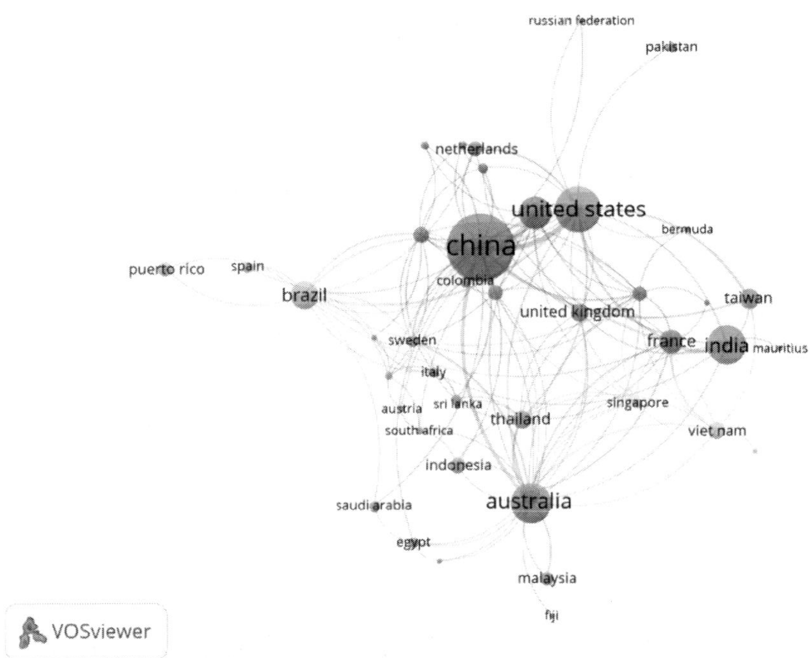

Figure 3. Analyze country cooperation networks that contribute to relevant publications.

Table 1. Top 20 productive countries related to microbial greenhouse gases in mangrove sediments

Ranking	Country	Documents	Citations	Total Link Strength	Average Citations
1	China	126	3659	54	29
2	United States	58	2879	43	50
3	Australia	45	2816	26	63
4	India	43	844	14	20
5	Hong Kong	29	1483	26	51
6	Brazil	23	979	11	43
7	France	16	558	22	35
8	Japan	12	470	14	39
9	Taiwan	12	341	8	28
10	Thailand	9	431	11	48
11	United Kingdom	9	292	10	32
12	Vietnam	8	213	11	27
13	Denmark	7	449	9	64
14	Germany	7	430	8	61
15	Indonesia	7	95	4	14
16	Canada	6	174	14	29
17	Mexico	6	320	3	53
18	Netherlands	6	164	6	27
19	Puerto Rico	6	331	1	55
20	Malaysia	5	359	3	72

Contribution of Related Institutions and Journal Sources

The results of the analysis using the VOS viewer reveal that 938 institutions play a role in contributing relevant publications. Table 2 shows the 10 highest affiliations in publications related to greenhouse gas microbes in mangrove sediments. On the basis of the ranking results, the three highest ranks, institutional affiliations that contribute to relevant publications, come from China. The affiliation with the most productive number of documents comes from the Ministry of Education of the People's Republic of China, with a total of 37 documents, followed by the University institution, Xiamen University, with 35 publications. These findings indicate that the trend of research on the role of microorganisms in the production of greenhouse gases is carried out in the mangrove ecosystem environment.

The results of the VOS viewer analysis revealed that 154 greenhouse gas microbes contributed to the mangrove sediments. Figure 4 shows the overlay of journals that published with "Science of the Total Environment" as the most influential source journal with the highest total citations and ahead of "Estuarine, Coastal and Shelf Science" in second place.

Table 2. Top ten most active institutions in the publication of relevant documents

Ranking	Institution	Documents
1	Ministry of Education of the People's Republic of China	37
2	Xiamen University	35
3	Chinese Academy of Sciences	28
4	The University of Hong Kong	18
5	Shenzhen University	13
6	National Institute of Oceanography India	13
7	Southern Cross University	13
8	State Key Laboratory of Marine Environmental Science	13
9	Australian Institute of Marine Science	12
10	Ministry of Natural Resources of the People's Republic of China	12

Table 3. Top 10 journal sources that publish greenhouse gas microbes in mangrove ecosystems

Ranking	Source	Documents	Citations	Total Link Strength	Average Citations
1	Science of the Total Environment	30	1208	39	40
2	Estuarine, Coastal and Shelf Science	17	1064	46	63
3	Marine Pollution Bulletin	17	596	28	35
4	Applied Microbiology and Biotechnology	13	787	28	61
5	Marine Ecology Progress Series	7	400	15	57
6	Biogeochemistry	6	316	16	53
7	Marine Environmental Research	6	115	27	19
8	Water Research	6	37	7	6
9	Plant and Soil	5	129	13	26
10	Soil Biology and Biochemistry	5	437	3	87

The top ten most fruitful journal sources concerning bacteria and greenhouse gases in mangrove sediments are displayed in Table 3. With a total of 30 papers and 1208 citations from Elsevier Publisher, the source publication "Science of the Total Environment" has the most documents. Furthermore, the second position is "estuarine, coastal and shelf science," with a total of 17 documents and 1064 citations, with the highest average total number of citations originating from Academic Press publishers. For the third rank, "Marine Pollution Bulletin" has 17documents with a total citation of 596 from the Elsevier publisher. All three journals are indexed by Scopus and ranked Q1.

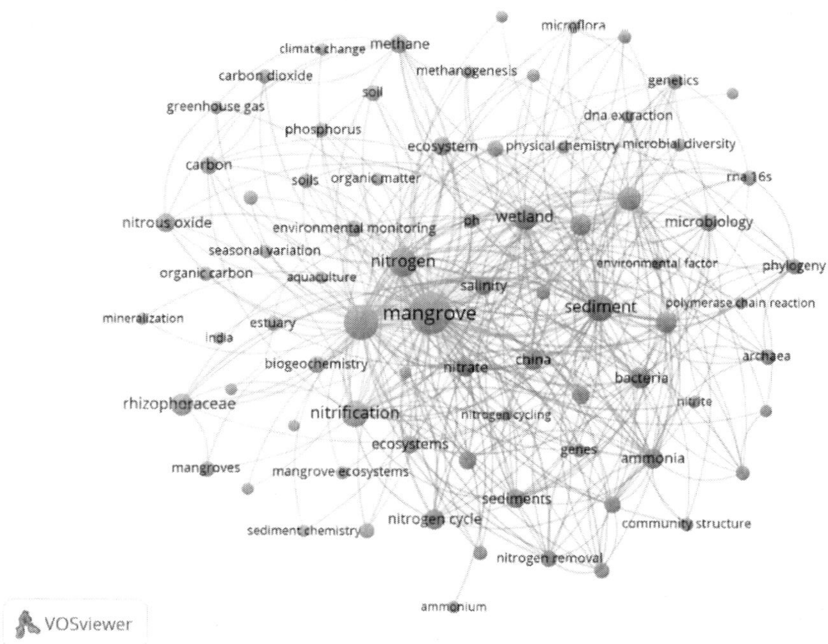

Figure 4. Keyword network visualization map based on co-occurrence.

Co-Occurrence Analysis of Keywords

The occurrence of relevant keywords was analyzed via VOSviewer. Keywords that had a strong relationship were grouped in each cluster. The cluster analysis results revealed 3 distinct clusters that addressed the role of microbes in the effects of greenhouse gases in mangrove sediments. Keyword mapping was built via relevant crucial keywords, and 83 keywords were found with each occurrence occurring at least 20 times in all the publications. These keywords were divided into three main clusters: cluster 1, red nodes focused on greenhouse gas and mangrove biogeochemistry; cluster 2, green nodes focused on nitrogen cycling and microbes; and cluster 3, blue nodes focused on microbiology and molecular methods (Figure 5).

Table 4. Keywords related to the greenhouse gas mechanism

Keyword	Occurrences
Mangrove	312
Denitrification	211
Nitrification	112
Methanogens	67
Anammox	36
Nitrogen Fixation	31
Methanotroph	26
DNRA	15
Comammox	2

Keyword analysis is performed to identify the topics that are relevant to the review conducted and predict the prospects of topics that will be discussed in the future. In addition, keyword analysis is needed to determine the relationship between each keyword. The terms that are most relevant to the mechanism of microbial greenhouse gases in mangrove sediments are displayed in Table 4. Denitrification has the most cooccurrences of any process, with 211, surpassing both nitrification and methanogens, which have 112 and 67 occurrences, respectively. It is critical to understand the kind of method that researchers have frequently employed in earlier research on the subject at hand.

Discussion

The bibliometric study yielded findings about study-related keywords and publication patterns, as well as organizations and journals that actively publish papers about bacteria that contribute to greenhouse gas emissions in mangrove sediments. To identify trends and research predictions that can be made to support the goals of the study, this analysis is crucial. The major greenhouse gases, such as N_2O, CH_4 and CO_2, are involved in the carbon and nitrogen cycles, most of which occur in mangrove sediments. The carbon and nitrogen cycles drive and regulate the occurrence of greenhouse gases, so scientific research is needed to mitigate the increase in greenhouse gas emissions and their contribution to climate change.

Carbon Cycle in Mangrove Ecosystems

The carbon cycle in mangrove ecosystems begins when mangroves absorb CO_2 through photosynthesis and store carbon in mangrove biomass (Basyuni et al., 2025). In addition, soil organic carbon can be formed from the activity of microorganisms in decomposing litter, breaking down organic matter into simpler compounds, and can store carbon in the form of microbial biomass (Sembiring and Sabrina, 2022).

The CO_2 compounds present in the soil can be utilized by methanogenic microorganisms in the methanogenesis mechanism through three main pathways. The three main methanogenesis pathways are hydrogenotrophic, acetoclastic, and methylotrophic. The first pathway is the hydrogenotrophic pathway, which uses hydrogen compounds (H_2) as electron donors alongside CO_2 as an electron acceptor to produce methane (CH_4). Examples of methanogenic bacteria include Methanococcales, Methanopyrales, and Methanobacteriales (Borrel et al., 2018). The second pathway is the acetoclastic pathway, which produces methane and CO_2 via acetate compounds, with examples of bacteria including *Methanosarcina* and*Methanosaeta* (Methanosarcinales) (Carr et al., 2018). The third pathway is the methylotrophic pathway, which uses H_2 to reduce methyl compounds and produce methane, with examples of microorganisms such as *Methanosphaera* and *Methanofastidiosa* (Borrel et al., 2014). On the basis of previous research, the dominant methanogenic bacteria in marine sediments are methylotrophs and acetoclastic bacteria, which contribute most significantly to methane release into the environment (Carr et al., 2018).

Methanotrophic bacterial populations have the ability to mitigate methane generation as a greenhouse gas in mangrove habitats. Methane can be broken down as a carbon source by methanotrophic bacteria, which then metabolize it to create CO_2 and biomass (Reis et al., 2022). Methanotrophs are divided into three types: Type I methanotrophs, which rely on the ribulose monophosphate pathway to assimilate carbon from methane via formaldehyde. Commonly found Type I methanotrophs include *Methylomonas, Methylomicrobium*, and *Methylococcus* (de la Torre, 2015). Next, type II methanotrophs are methanotrophs that rely on the seryl-CoA cycle to oxidize methane. These methanotrophs tend to be found in methane-rich conditions. Examples of bacteria involved in methane oxidation of type II include *Methylocystis* and *Methylosinus* (Ahmadi and Lackner, 2024). Methanotroph type X has the characteristic of ribulose-1,5-bisphosphate carboxylase and uses the enzymes pMMO (particulate methane monooxygenase) and sMMO (soluble methane monooxygenase) in methane oxidation (Knief, 2015). Type I methanotrophs, which oxidize methane, are the most abundant type in mangrove ecosystems and have a greater tolerance to salinity than other types do, indicating a greater level of adaptation (Shiau et al., 2018). This study revealed that the number of cooccurrences of methanogenic stages was greater (67) than that of methanotrophic stages (26), which play a role in the carbon cycle in mangrove sediments.

The Nitrogen Cycle in Mangrove Ecosystems

As a biogeochemical cycle, the nitrogen cycle explains how nitrogen ions and molecules change via a variety of processes. Mangrove habitats have the capacity to change nitrogen through the action of microorganisms (Mubaraq et al., 2025). Excessive nitrogen accumulation in mangrove sediments can pose a threat to biodiversity. Excess N supports biological invasion and competition between species, leading to changes in species dominance patterns in an environment and suppressing species biodiversity and the soil biota (Reis et al., 2017).

The emission of greenhouse gases is significantly influenced by the nitrogen cycle and microbial interactions. There are three main pathways involved in nitrogen transformation in the environment: N_2 fixation, nitrification, and denitrification (Han et al., 2024). The initial stage of the nitrogen cycle is nitrogen fixation, which is the process of absorbing N_2 from the atmosphere and converting it into ammonia (NH_3) via enzymatic reactions involving nitrogenase enzymes, such as those found in bacteria such as *Azotobacter* and *Rhizobium* (Shiau et al., 2017). Additionally, organic nitrogen compounds from other organisms, such as animals and plants, are decomposed into ammonia and ammonium in mangrove sediments through ammonification. Several types of ammonifying bacteria, including *Bacillus, Clostridium, Pseudomonas*, and *Streptomyces, have been identified in mangrove sediments* (Dai et al., 2021).

Ammonium compounds derived from nitrogen fixation and ammonification activities are converted into nitrite and nitrate compounds through the nitrification process, which is an important step in the global nitrogen cycle (Chen et al., 2015). The nitrification process occurs in two stages: the conversion of ammonium to nitrite with hydroxylamine as an intermediate by ammonium-oxidizing microorganisms via the enzymes ammonia monooxygenase and hydroxylamine oxidoreductase. Some types of ammonium-oxidizing bacteria belong to the genera *Nitrosomonas, Nitrosospira,* and *Nitrosococcus*, and this process can also be carried out by archaea of the genus Nitrosopumilus. The second stage involves the conversion of nitrite (NO_2-) to nitrate (NO_3-) by nitrite-oxidizing bacteria (NOB) via the enzyme nitrite oxidoreductase (Wang et al., 2015). To date, bacteria capable of converting ammonium to nitrate within a single microorganism, such as the Comammox bacteria within the genus *Nitrospira, have been identified* (Liu et al., 2020).

Nitrate present in mangrove sediments can then be reduced to nitrogen gas (N_2) via intermediate compounds such as nitrite, nitric oxide (NO), nitrous oxide (N_2O), and finally nitrogen gas through the denitrification mechanism. Denitrification can be a key process in nitrogen removal in mangrove sediments carried out by bacteria such as *Pseudomonas*, *Alcaligenes*, and *Paracoccus* (Dai et al., 2021). Nitrate can also be reduced to ammonium through dissimilatory nitrate reduction to ammonium (DNRA).

Conclusion

Mangrove ecosystems, which are places where land and sea meet, are home to a variety of microorganisms that contribute to the production and control of greenhouse gases along coasts. To identify publishing trends, patterns, and connections among the scientific literature on microorganisms contributing to greenhouse gas emissions in mangrove sediments, bibliometric analysis was employed. This bibliometric analysis can be used as a reference for mapping future research. The results of the bibliometric analysis revealed that among the key mechanisms involved in greenhouse gas mechanisms in mangrove ecosystems, denitrification is the most widely studied mechanism. Denitrification can be investigated in the use of microbes in mangrove habitats to remove nitrogen, which is then released into the atmosphere as N_2. To understand the carbon and nitrogen management mechanisms in mangrove ecosystems, the greenhouse gas mechanism has been studied.

Disclaimer

None.

References

Ahmadi, F., Lackner, M. 2024. Recent findings in methanotrophs: genetics, molecular ecology, and biopotential. *Applied Microbiology and Biotechnology*, 108(1): 60.

Basyuni, M., Mubaraq, A., Amelia, R., Wirasatriya, A., Iryanthony, S. B., Slamet, B., Arifanti, V. B. 2025. Mangrove aboveground biomass estimation using UAV imagery and a constructed height model in Budeng–Perancak, Bali, Indonesia. *Ecological Informatics*, 86: 103037.

Borrel, G., Adam, P. S., McKay, L. J., Chen, L. X., Sierra-García, I. N., Sieber, C. M., Gribaldo, S. 2019. Wide diversity of methane and short-chain alkane metabolisms in uncultured archaea. *Nature microbiology*, 4(4): 603-613.

Borrel, G., Gaci, N., Peyret, P., O Toole, P. W., Gribaldo, S., Brugère, J. F. 2014. Unique characteristics of the pyrrolysine system in the 7th order of methanogens: implications for the evolution of a genetic code expansion cassette. *Archaea*, 2014(1), 374146.

Carr, S. A., Schubotz, F., Dunbar, R. B., Mills, C. T., Dias, R., Summons, R. E., Mandernack, K. W. 2018. Acetoclastic *Methanosaeta* are dominant methanogens in organic-rich Antarctic marine sediments. *The ISME Journal*, 12(2): 330-342.

Chen, J., Zhou, H. C., Pan, Y., Shyla, F. S., Tam, N. F. Y. 2016. Effects of polybrominated diphenyl ethers and plant species on nitrification, denitrification and anammox in mangrove soils. *Science of the Total Environment*, 553: 60-70.

Dai, Y., Lin, X., Luo, Y., Sun, J., Tian, Y. 2021. Molecular analysis of microbial nitrogen transformation and removal potential in mangrove wetlands under anthropogenic nitrogen input. *Science of The Total Environment*, 773: 145632.

Das, N., Mandal, S. 2022. Microbial populations regulate greenhouse gas emissions in Sundarban mangrove ecosystem, India. *Acta Ecologica Sinica*, 42(6): 641-652.

de la Torre, A., Metivier, A., Chu, F., Laurens, L. M., Beck, D. A., Pienkos, P. T., Kalyuzhnaya, M. G. 2015 Genome-scale metabolic reconstructions and theoretical investigation of methane conversion in *Methylomicrobium buryatense* strain 5G (B1). *Microbial cell factories*, 14(1): 188.

Donthu, N., Kumar, S., Mukherjee, D., Pandey, N., Lim, W. M. 2021. How to conduct a bibliometric analysis: An overview and guidelines. *Journal of business research*, 133: 285-296.

Ellegaard, O., Wallin, J. A. 2015. The bibliometric analysis of scholarly production: How great is the impact? *Scientometrics*, 105(3); 1809-1831.

Ghizelini, A. M., Mendonça-Hagler, L. C. S., & Macrae, A. 2012. Microbial diversity in Brazilian mangrove sediments: a mini review. *Brazilian Journal of Microbiology*, 43: 1242-1254.

Han, P., Tang, X., Koch, H., Dong, X., Hou, L., Wang, D., Shi, G. 2024. Unveiling unique microbial nitrogen cycling and nitrification driver in coastal Antarctica. *Nature Communications*, 15(1): 3143.

Holguin, G., Gonzalez-Zamorano, P., De-Bashan, L. E., Mendoza, R., Amador, E., Bashan, Y. 2006. Mangrove health in an arid environment encroached by urban development—a case study. *Science of the Total Environment*, 363(1-3): 260-274.

Knief, C. 2015. Diversity and habitat preferences of cultivated and uncultivated aerobic methanotrophic bacteria evaluated based on pmoA as molecular marker. *Frontiers in microbiology*, 6: 1346.

Liao, H., Tang, M., Luo, L., Li, C., Chiclana, F., Zeng, X. J. 2018. A bibliometric analysis and visualization of medical big data research. *Sustainability*, 10(1): 166.

Lin, X., Zhu, B., Peng, Y., Lin, G., Sun, D., & Ye, W. 2025. Responses of microbial communities and greenhouse gas production to land use change in mangrove wetland sediments. *Limnology and Oceanography*, 70(5): 1210-1225.

Liu, Z., Zhang, C., Wei, Q., Zhang, S., Quan, Z., Li, M. 2020. Temperature and salinity drive comammox community composition in mangrove ecosystems across southeastern China. *Science of The Total Environment*, 742: 140456.

Ma, Y., Wang, W., Gao, F., Yu, C., Feng, Y., Gao, L., Xie, J. 2024. Acidification and hypoxia in seawater, and pollutant enrichment in the sediments of qiao island mangrove wetlands, pearl river estuary, China. *Ecological Indicators*, 158: 111589.

Mayer, F., Enzmann, F., Lopez, A. M., Holtmann, D. 2019. Performance of different methanogenic species for the microbial electrosynthesis of methane from carbon dioxide. *Bioresource technology*, 289: 121706.

Mubaraq, A., Basyuni, M., Sembiring, M., Aznawi, A. A., Sivaipram, I. 2025. Bioinformatics approach of nitrifying microorganism genes from mangrove ecosystems. In *IOP Conference Series: Earth and Environmental Science* 1445(1): 012119.

Mubaraq, A., Sembiring, M., Widiastuti, E., Fachrial, E., Utomo, B., Turjaman, M., Basyuni, M. 2024. Diversity and distribution of nitrifying bacteria play an important role in the nitrogen cycle in mangrove sediments. *Global Journal of Environmental Science & Management (GJESM)*, 10(4).

Qian, Z., Li, Y., Pratush, A., Kan, J., Gu, J. D., Peng, T., Hu, Z. 2024. A comparative analysis of the microbial communities and functional genes of the nitrogen cycling in mangroves of China, Indian and Malaysia. *International Biodeterioration & Biodegradation*, 190: 105767.

Reis, C. R. G., Nardoto, G. B., Oliveira, R. S. 2017. Global overview on nitrogen dynamics in mangroves and consequences of increasing nitrogen availability for these systems. *Plant and soil*, 410(1): 1-19.

Reis, P. C., Thottathil, S. D., Prairie, Y. T. 2022. The role of methanotrophy in the microbial carbon metabolism of temperate lakes. *Nature Communications*, 13(1): 43.

Rosentreter, J. A., Maher, D. T., Erler, D. V., Murray, R. H., Eyre, B. D. 2018. Methane emissions partially offset "blue carbon" burial in mangroves. *Science advances*, 4(6): 4985.

Sembiring, M., Sabrina, T. 2022. Diversity of potassium solving microbes on andisol soil affected by the eruption of Mount Sinabung, North Sumatra, Indonesia. *Biodiversitas Journal of Biological Diversity,* 23(4).

Sengupta, T., Mitra, A. 2020. Spatial variation of the microbial diversity in the mangrove dominated Sundarban Forest of India. In *Recent Advancements in Microbial Diversity* (pp. 333-350).

Shiau, Y. J., Cai, Y., Lin, Y. T., Jia, Z., Chiu, C. Y. 2018. Community structure of active aerobic methanotrophs in red mangrove (*Kandelia obovata*) soils under different frequency of tides. *Microbial Ecology*, 75(3): 761-770.

Shiau, Y. J., Chiu, C. Y. 2020. Biogeochemical processes of C and N in the soil of mangrove forest ecosystems. *Forests*, 11(5): 492.

Shiau, Y. J., Lin, M. F., Tan, C. C., Tian, G., & Chiu, C. Y. 2017. Assessing N_2 fixation in estuarine mangrove soils. *Estuarine, Coastal and Shelf Science*, 189: 84-89.

Thatoi, H., Behera, B. C., Mishra, R. R., Dutta, S. K. 2013. Biodiversity and biotechnological potential of microorganisms from mangrove ecosystems: a review. *Annals of Microbiology*, 63(1): 1-19.

Van Eck, N., Waltman, L. 2010. Software survey: VOSviewer, a computer program for bibliometric mapping. *scientometrics*, 84(2): 523-538.

Wang, H., Su, J., Zheng, T., Yang, X. 2015. Insights into the role of plant on ammonia-oxidizing bacteria and archaea in the mangrove ecosystem. *Journal of Soils and Sediments*, 15(5): 1212-1223.

Yang, Y., Wang, W., Hu, Q., Yao, X., Yang, W., Wen, S., Shen, L. 2025. Conversion of coastal wetlands to paddy fields substantially decreases methane oxidation potential and methanotrophic abundance on the eastern coast of China. *Water Research*, 272: 122962.

Chapter 5

The Silent Forest: Trade-Offs in Mangrove Restoration for Climate and Biodiversity Goals

Suyadi[1,*]
Yulizar Ihrami Rahmila[1]
Muhammad Faisal Rachmansyah[2]
Muksin Purnama[2]
Marwan Setiawan[1]
Mira Yulianti[1]
Varenna Faubiany[1]
Zilda Dona Okta Permata[1]
Parlin Hotmartua Putra Pasaribu[1]
Hedi Indra Januar[1]
Rinny Rahmania[1]
Izhamil Hidayah[1]
Yaya Ihya Ulumudin[1]
Bayu Prayudha[1]
and Restu Dewinta Maharani[3]
[1]Research Center for Ecoloy, National Research and Innovation Agency, Bogor, Indonesia
[2]Marine and Fisheries Department, Diponegoro University, Semarang, Indonesia
[3]Environment Management Department, Pakuan University, Bogor, Indonesia

---

[*] Corresponding Author's Email: suyadipedhet@gmail.com

In: Biodiversity
Editors: Mohammad Basyuni and Faujiah Nurhasanah Ritonga
ISBN: 979-8-90134-246-6

Abstract

Mangrove planting is the most common approach used in many countries, including Indonesia, to increase mangrove areas. A large amount of funding and effort has been invested in mangrove planting, but the success rate in terms of seed/propagule survival and long-term mangrove area increase is relatively low. This study investigated the long journey of mangrove restoration in Indonesia and evaluated planted mangroves in terms of species diversity (mollusks and crabs) and carbon storage. Most mangrove plants in Indonesia use a single species of mangrove (monocultures). Compared with naturally growing mangroves, planted mangroves present significantly greater tree density (18,485 ± 628 stem/ha, χ^2 (1) = 19.2, $p < 0.000$). and canopy coverage but lower in diameter, tree height, species diversity, and seedling regeneration. The aboveground carbon of planted mangroves (34.61 Mg ha^{-1}) was significantly greater (up to 230.5%) than that of naturally growing mangroves (8.95 Mg ha^{-1}), χ^2 (1) = 18.8, $p < 0.000$. However, the predominance of high-density but small-diameter trees suggests that carbon storage may plateau or decline as the forest structure matures. In addition, the species diversity of mollusks in planted mangroves was significantly lower (up to 92%) than that in naturally growing mangroves, and the species diversity of crabs in planted mangroves was much lower (up to 95%) than that in naturally growing mangroves. This study demonstrated that single-species plantations, although efficient for rapid greening and carbon sequestration, are insufficient for restoring the full ecological functions of mangrove ecosystems. Restoration projects should prioritize not only carbon targets but also multispecies planting and habitat complexity to foster faunal diversity and broader ecological functioning.

Keywords: mangroves, carbon storage, faunal diversity, carbon and biodiversity imbalance

Introduction

Indonesia is a global center of mangrove biodiversity, hosting the highest number of true mangrove species (45 species, 60%) in the world (Giesen et al., 2007). Both true and associated mangroves can be found in Indonesia. Mangrove forests in Indonesia are distributed along the coastlines of nearly all major islands, with high concentrations in Papua, Kalimantan, Sumatra, Sulawesi, and Java. According to the National Mangrove Map 2021, Indonesia provides habitat for approximately 3.36 million hectares of mangroves, accounting for more than 23% of the world's mangrove area (MoEF, 2021).

Mangroves provide critical ecological services that protect both marine and terrestrial environments. They function as natural coastal buffers, reducing erosion and attenuating wave energy, thereby protecting shorelines from storms and sea-level rise (Alongi, 2008). Mangrove ecosystems not only support rich flora but also host numerous faunal species, including fish, crustaceans, mollusks, reptiles, and birds, many of which are economically and ecologically important (Alongi, 2002). Furthermore, mangroves are highly efficient carbon sinks; they store significant amounts of carbon both above and below ground, making them vital ecosystems for climate change mitigation (Donato et al., 2011).

Economically, mangrove forests support livelihoods and national economies through fisheries, aquaculture, forestry, and ecotourism. Mangrove wood and nontimber products (e.g., honey, food, medicinal plants) are also important sources for local communities. In monetary terms, the ecosystem services provided by mangroves are valued at thousands of dollars per hectare per year, with coastal protection alone estimated at over US$194,000 per hectare to avoid damage from storm surges and erosion (Barbier et al., 2011). Additionally, mangroves play a growing role in carbon markets through initiatives such as blue carbon, where carbon sequestration by mangroves is monetized to support conservation and community development.

Despite their ecological and economic value, Indonesian mangroves are under significant threat. Historically, Indonesia lost more than 40% of its mangrove cover between 1980 and 2005, primarily because of the conversion of mangrove forests into shrimp ponds, agriculture, and settlements (FAO, 2007; Murdiyarso et al., 2015). Moreover, legal and illegal logging for fuelwood and timber further exacerbates degradation.

This destruction not only reduces biodiversity but also weakens coastal resilience against climate change impacts such as sea-level rise, erosion, and storm surges (Alongi, 2002). Hydrological alterations, such as canal construction and road building, hinder tidal exchange and salinity balance, severely affecting species composition and productivity. Furthermore, degraded mangroves release stored carbon into the atmosphere, contributing to greenhouse gas emissions (Donato et al., 2011).

In recent years, national and international efforts have sought to reverse this trend through restoration and rehabilitation initiatives. This initiative is aligned with broader climate change mitigation goals, including the Nationally Determined Contributions (NDCs) under the Paris Agreement. Mangrove ecosystems are among the most carbon-rich forests in the tropics, storing up to 1,000 Mg C ha^{-1} in their biomass and soils, with the majority sequestered below ground (Donato et al., 2011). The restoration of degraded mangroves thus presents a substantial opportunity for enhancing "blue carbon" stocks and mitigating climate change. These factors not only help in recovering biodiversity and ecosystem services but also create opportunities to participate in emerging carbon markets.

However, challenges remain, including programs and funding mechanisms, which have incentivized mangrove restoration and rehabilitation (mangrove planting) with a strong emphasis on quantifiable carbon and biodiversity outcomes (Murdiyarso et al., 2015; Howard et al., 2017; Suyadi et al., 2023). Data and scientific information highlight that the unbalanced attention given to the purposes of mangrove restoration and rehabilitation (carbon and biodiversity) is still limited. The aim of this study was to assess the trends in the purposes of mangrove restoration and rehabilitation and their consequences for carbon and biodiversity. To achieve these objectives, several methods were applied: systematic literature review, vegetation transects/plots, and fauna (mollusk and crabs) data and sample collection. This study describes the trend of mangrove restoration and rehabilitation and highlights the importance of balancing attention to carbon and biodiversity. This information is required to place mangrove restoration and rehabilitation activities as the main goals of biodiversity conservation and climate mitigation and adaptation.

Methods

The study was conducted in Indonesia, in which vegetation and fauna (mollusk and crab) data and samples were collected in Segara Anakan, Cilacap, Central Java and Segara Anakan, Banyuwangi, East Java. These research locations were selected because of their similar environments (lagoon), and both planted and naturally growing mangroves exist in these lagoons. The mangrove ecosystems in Segara Anakan, Cilacap and Segara Anakan, Banyuwangi are characterized as estuarine mangroves and terrigenous sediment, which is typically dominated by clays and loam within highly silted mud substrates.

Segara Anakan, Cilacap (7°43'48.2"S 108°59'12.1"E and 7°40'34.2"S 108°48'25.5"E), is the largest mangrove habitat in Java. The water salinity at this study site generally falls within 0–40 ppt, while the measured surface salinity is often between 0 and 7.05 ppt, the water pH is between 5.6 and 7.1, and the soil pH is between 5.7 and 6.9. The water temperature in the area ranges from approximately 26.5°C to 28.3°C, with high humidity (81.9–87%) and wind speed fluctuations between 2.3 and 5.9 knots; rainfall is highly variable, with extremes ranging from 0.2 mm to nearly 800 mm monthly, influencing freshwater inflow and salinity gradients. Geographically, Segara Anakan, Banyuwangi, East Java, is located between 8°35'50.7"S 114°13'40.4"E and 8°37'34.0"S 114°20'39.7"E. The temperature range in the mangrove ecosystem is between 23°C and 33°C, with a pH of approximately 7.0 and salinity values between 23‰ and ≤36‰.

This research applied three main methods: literature review, vegetation transects/plots, and faunal (mollusk and crab) samples collected from plots/quadrants (Figure 1). The literature review focused on planted and naturally growing mangroves, carbon stocks, and biodiversity. Data and information were collected from published literature, unpublished literature such as reports from the last decade (1923–2024), and the official website of mangrove planting: https://www.padatkaryamangrove.info. The relevant literature was searched via the search engines Scopus, ISI Web of Science (Thomson Reuters), and Google Scholar, which employ a combination of specific keywords: "mangroves," "planting," "restoration," "rehabilitation," "carbon stock," "biodiversity," "crabs," and "mollusk."

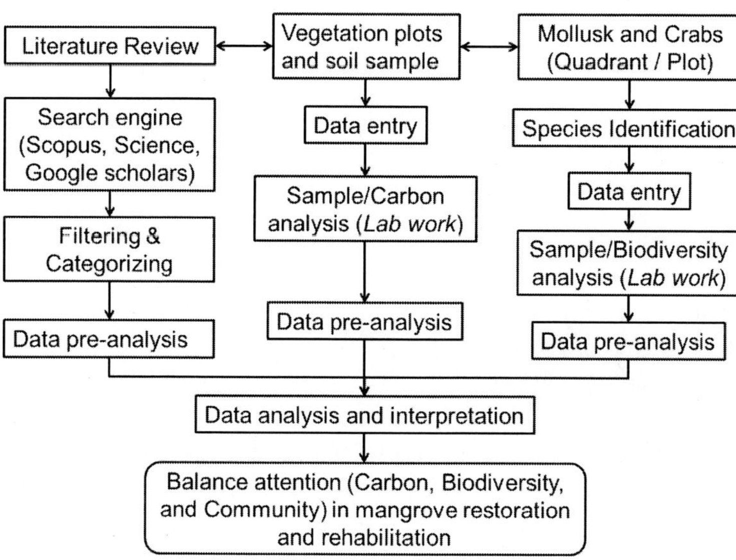

Figure 1. Flowchart of methods (literature review, vegetation plots, and the Mollusk and Crabs quadrant) applied to achieve the objectives of the research.

Papers published in journals and proceedings (international, national, and local) and graduate theses were included in the study. A manual search was applied to search for papers that were not published online. The next step was filtering, which was performed manually to ensure that the literature was relevant to the topics and was included in the next steps. In this step, all the papers were fully read and summarized. The relevant papers and the summary were then categorized according to the main topics and subtopics of this report. Finally, data extraction and analysis were conducted to obtain numeric data and descriptive information.

Vegetation data were collected from 10 × 10 m plots systematically distributed across the study areas to represent the variation in mangrove communities in Segara Anakan, Cilacap (38 sampling points) and Segara Anakan, Banyuwangi (12 sampling points). The sampling points were determined via a stratified random sampling method with proportional area allocation (overall expected standard error: 0.01 or 1%). All trees with a diameter greater than 5 cm within each plot were identified to the species level and measured for their diameter 134 cm above the ground surface, corresponding to the diameter at breast height (DBH). Saplings (DBH 2–4.9 cm) within the plots were also identified and measured for diameter. For nipa palms (Nypa fruticans), the diameter of the frond petiole was measured (Rahman et al., 2020). Moreover, the seedlings (diameter < 2 cm) present in the 1 m^2 quadrants within the plots were identified to the species level and counted.

Mollusk samples were collected from 1 × 1 meter quadrants located in the plot, including individuals found on the land surface, mangroves up to 10 cm from the land surface, and within the soil layer up to a depth of 10 cm. Crab samples were collected by a single observer within a 10 × 10 meter plot, with a sampling duration of one hour per plot. The mollusk and crab samples were subsequently cleaned and preserved in alcohol before being transported to the laboratory for species identification.

Data derived from the literature were analyzed descriptively to determine the extent of the land planted with mangroves, the number of mangroves planted, the planting approaches and techniques applied, and the species of mangroves planted. The vegetation data were analyzed to assess the structure and composition of the mangrove vegetation, tree density, canopy cover, vegetation conditions, and carbon stock. Mollusk and crab data were analyzed to determine population density and species diversity.

Prior to analysis, the data were tested for normality via the Shapiro–Wilk test. The Kruskal–Wallis test, followed by a Bonferroni post hoc correction, was applied to determine if there were significant differences in parameters between naturally growing and planted mangroves. Linear regression analysis was used to assess the relationship between the DBH of trees and tree density. Wilcoxon tests were applied to compare carbon stocks in naturally growing and planted mangroves. Kruskal–Wallis, linear regression, and Wilcoxon tests were performed via SPSS 22 (IBM, New York, USA).

Results

Mangrove Planting Techniques

The most popular mangrove planting approach applied in Indonesia is ecological mangrove restoration (EMR), with some modifications depending on the extent of degradation, environmental conditions, local communities, and available resources. Minor and medium hydrological modifications are common land preparation techniques applied in Indonesia, but many mangrove plantings have been performed without or with less proper land preparation. In addition, the role of the local community in mangrove planting is crucial for all aspects and technical steps from the mangrove nursery until plant care and monitoring.

Our observations and literature review revealed that at least eight land preparation techniques have been applied in Indonesia: the *cemplongan* technique, the water break technique, the bamboo pole technique, the *guludan* technique, the enormous polybag technique, the ditch muddy technique, the cluster technique, and the enormous hole muddy technique. In addition, there are several mangrove planting patterns/designs, such as square planting design, zig-zag planting design, and cluster planting design. The application of these mangrove planting designs depends on the condition of the land, the purpose of mangrove planting, and human resources.

Active planting (plant seedlings or propagules) is more popular in Indonesia than natural regeneration. The distance most commonly applied among seedlings is 1 × 1 m. Other planting distances are 1 × 2 m, 1.5 × 1.5 m, 2 × 2 m, and 2 × 3 m. Another mangrove planting design is the stripe mangrove planting design, which is usually applied in areas dominated by other plants, such as *Nypa frutic*ans and *Derris trifoliata*. A combination of a mangrove planting technique and a mangrove planting design is needed to improve the success of mangrove planting.

Most mangrove plants in Indonesia use a single species of mangrove (monocultures). Data derived from BRGM (https://www.padatkarya mangrove.info) revealed that mangrove planting in Java between 2021 and 2021 was conducted in 224 locations (total area 3315, 56 hectares). The most common species of mangroves planted are *Rhizophora apiculata* and *Rhizophora mucronata* (Table 1). These are generalist species that can be found in many locations, and the population density of these species is relatively high. In addition, propagules of these species are naturally available throughout the year. In addition, the propagules of *Rhizophora apiculata* and *Rhizophora mucronata* are relatively easy to transport, and their viability is high. These species are also known as pioneer species growing in fringe areas (seaward zones). *Avicennia marina* and *Avicennia officinalis* are also planted by local communities, accounting for 5.93% and 5.53%, respectively (Table 1). These species are also distributed across Indonesia, but their fruits are seasonal.

Table 1. List of species commonly planted by local community groups from 2020-2021

No	Family	Species	Number of planted locations (%)	Planted area (%)
1	RHIZOPHORACEAE	*Rhizophora apiculata*	48,22	57,47
2	RHIZOPHORACEAE	*Rhizophora mucronata*	29,64	28,69
3	AVICENNIACEAE	*Avicennia marina*	5,93	3,23
4	AVICENNIACEAE	*Avicennia officinalis*	5,53	3,09
5	RHIZOPHORACEAE	*Rhizophora stylosa*	4,74	3,62
6	RHIZOPHORACEAE	*Bruguiera gymnorrhiza*	3,16	1,78
7	SONNERATIACEAE	*Sonneratia alba*	1,98	1,66
8	RHIZOPHORACEAE	*Ceriops decandra*	0,40	0,24
9	RHIZOPHORACEAE	*Ceriops tagal*	0,40	0,21
Total number of planted location = 224 locations				
Total planted area = 3315,56 hectares				

Vegetation Characteristics of Natural-Growth and Planted Mangroves

Compared with their natural counterparts, planted mangroves in both Banyuwangi and Cilacap demonstrate significantly greater tree density (Table 2). For example, in Banyuwangi, the tree density of planted mangroves reaches 18,485 ± 628 stem/ha, which is more than double that of naturally growing mangroves (8043 ± 365 stem/ha, χ^2 (1) = 19.2, p< 0.000). A similar trend was observed in Cilacap, although to a lesser extent (6978 ± 424 vs. 5978 ± 261 stem/ha, χ^2 (1) = 18.7, p< 0.000).

Table 2. Vegetation characteristics of natural-growth and planted mangroves

Parameters	Banyuwangi		Cilacap	
	Natural growth	Planted mangrove	Natural growth	Planted mangrove
Tree density (stem/ha)	8043 ± 365	18485 ± 628	5978 ± 261	6978 ± 424
Canopy coverage (%)	69,69 ± 1,33	72,25 ± 1,54	70,85 ± 2,14	77,60 ± 1,97
Mean DBH (cm)	10,72 ± 1,96	4,38 ± 1,58	6,63 ± 1,42	4,41 ± 1,32
Mean tree high (m)	9,46 ± 0,43	5,01 ± 0,41	8,56 ± 0,33	5,67 ± 0,19
Seedling (stem/ha)	3203 ± 418	2193 ± 645	1084 ± 172	883 ± 135
Simpson's diversity index	5,86	2,19	6,47	2,26
Conditions*	Dense	Dense	Dense	Very dense

*Kepmen Lingkungan No. 201 2004.

However, planted mangroves consistently have lower mean DBH and tree height values than naturally growing mangroves do. In Banyuwangi, natural mangroves have a mean DBH of 10.72 ± 1.96 cm and a tree height of 9.46 ± 0.43 m, whereas planted stands measure only 4.38 ± 1.58 cm, χ^2 (1) = 16.3, p< 0.000 and 5.01 ± 0.41 m, χ^2 (1) = 17.8, p< 0.000, respectively. In Cilacap, a similar pattern is evident, with planted mangroves having smaller DBHs and shorter heights. In addition, naturally growing mangroves presented significantly greater Simpson's diversity index values, with Cilacap's natural stands scoring 6.47 and Banyuwangi's scoring 5.86. In contrast, planted mangroves in the same region score much lower (2.26 and 2.19, χ^2 (1) = 19.5, p< 0.000, respectively).

Data analysis revealed significant structural and compositional differences between naturally growing mangroves and planted mangroves in both locations (Banyuwangi and Cilacap), which provides insight into forest maturity, health, and biodiversity levels. Following the mangrove conditions standard described by the Ministry of the Environment of the Republic of Indonesia No 201 Year 2004 (Kepmen Lingkungan No. 201 Tahun 2004), both the natural-growth and planted mangrove areas of Banyuwangi and Cilacap are in good condition, with the planted mangroves in Cilacap categorized as very dense mangroves. Overall, the data suggest that while planted mangroves can achieve high tree density and canopy coverage, they generally exhibit lower biodiversity and structural maturity (in terms of mean DBH, regeneration/seedlings) than naturally grown mangrove stands do.

Carbon Stock of Natural-Growth and Planted Mangroves

Carbon analysis revealed that, compared with naturally growing mangroves, planted mangroves store more total carbon, primarily due to differences in aboveground carbon (AGC) and root belowground carbon (BGC) components (Figure 2). The aboveground carbon of the planted mangroves in Banyuwangi and Cilacap is approximately 230.5% and 182.4% greater than that of the naturally growing mangroves, $\chi^2(1) = 18.8, p < 0.000$, respectively. Similarly, the amount of root belowground carbon in planted mangroves is 5.31 to 6.03% greater than that associated with natural growth. In contrast, the soil carbon content of the planted mangrove plants in both Banyuwangi and Cilacap was lower than that of the naturally growing plants, at 5.4% and 6.1%, respectively.

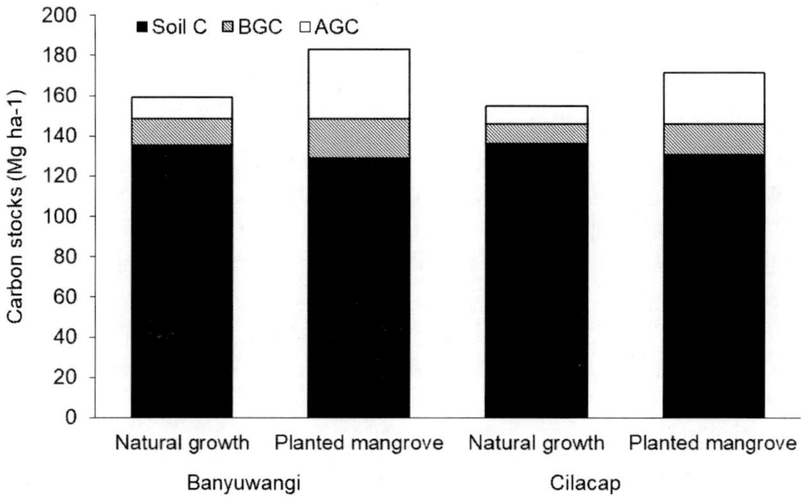

Figure 2. Carbon stocks (soil carbon, belowground carbon/roots, and aboveground carbon) of naturally growing and planted mangrove plants in Banyuwangi and Cilacap.

Compared with naturally growing mangroves, planted mangroves store more carbon because the density of trees is high but is dominated by relatively small DBH trees (Table 2). This result was supported by correlation analysis, which revealed that DBH significantly decreased with increasing tree density in both Banyuwangi ($r^2 = 0.83, p < 0.001$) and Cilacap ($r^2 = 0.78, p < 0.001$) trees (Figure 3). This means that the carbon stock in planted mangroves results from the high density of small mangrove stems rather than large stems. This is in line with vegetation characteristics, in which the tree density and canopy coverage of planted mangroves are high, but the DBH, tree height, and species diversity of mangroves are low. In addition, the regeneration (seedling density) of planted mangroves is much lower than that of naturally growing mangroves.

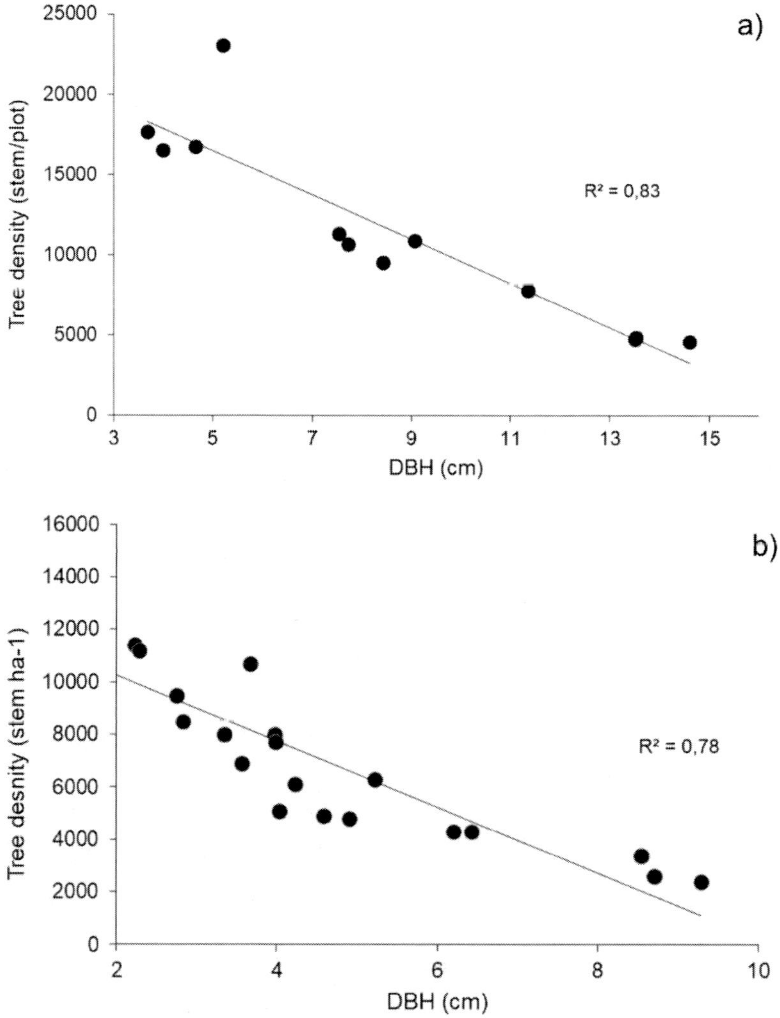

Figure 3. Correlation of tree density and tree diameter as a basis of carbon calculation in the mangrove ecosystems of Banyuwangi a) and Cilacap b).

Species Diversity of Mollusks and Crabs in Planted Mangroves and Natural-Growth Mangroves

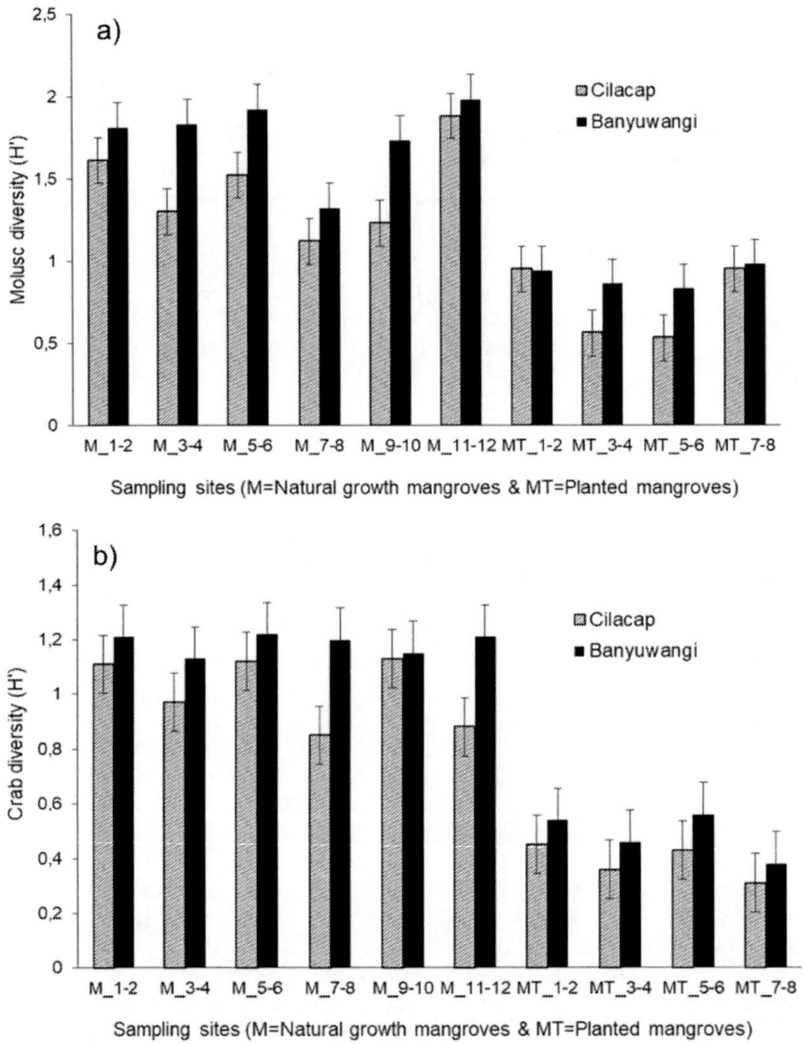

Figure 4. Diversity of moluscs (a) and crabs (b) in Cilacap and Banyuwangi.

This study revealed 26 mollusk species and 12 crab species in the study area. The most dominant mollusk is *Neripteron violaceum,* followed by *Littoraria carinifera, Cassidula aurisfelis*, and *Nerita balteata*. The dominant species of crabs is *Austrucas* spp., followed by *Parasesarma* sp., *Metaplax elegans*, and *Episesarma lafondi.* The population of mollusks and crabs in planted mangroves was significantly lower (up to 131 individuals/plot from 9 species) than that in naturally growing mangroves (up to 282 individuals/plot from 16 species), χ^2 (1) = 16.9, $p< 0.000$.

The species diversity of mollusks in planted Cilacap mangroves was significantly lower (mean diversity index: 0.75) than that in naturally growing mangroves (mean diversity index: 1.44), χ^2 (1) = 18.2, $p< 0.000$ (Figure 4). Like Cilacap, the species diversity of mollusks in the planted mangrove community of Banyuwangi was significantly lower (mean diversity index: 0.90) than that in the naturally growing mangrove community (mean diversity index: 1.76), χ^2 (1) = 17.6, $p< 0.000$. Similarly, the species diversity of crabs in planted mangroves was also much lower (mean diversity index: 0.48) than that in naturally growing mangroves (mean diversity index: 1.01) for Cilacap, χ^2 (1) = 19.1, $p< 0.000$. The species diversity of crabs in the planted mangrove stands of Banyuwangi was significantly lower (mean diversity index: 0.90) than that in the naturally growing mangroves (mean diversity index: 1.76), χ^2 (1) = 17.9, $p< 0.000$.

Discussion

Carbon Stock and Faunal Diversity in Planted vs. Natural Mangrove Forests

Compared with naturally growing mangroves, planted mangroves store significantly more aboveground carbon (AGC), which can be attributed primarily to the high stem density of small mangrove stems in planted mangroves. The density of planted mangrove stands was more than twice that of natural growth forests in some cases, despite having smaller mean diameters and shorter heights. This structural configuration results in greater cumulative biomass per hectare, especially in AGC, as many small stems can collectively surpass the biomass of fewer large trees (Alongi, 2012; Murdiyarso et al., 2015). Furthermore, the amount of root belowground carbon (BGC) was also slightly greater in planted mangroves than in cultivated ones, likely because of the dense root systems and proximity of planting, which increased the fine root biomass.

However, despite greater carbon storage, planted mangroves consistently presented lower species diversity of both mollusks and crabs. This pattern is likely due to the simplified structural complexity and species composition typical of restoration projects, which often use monoculture or low-diversity planting schemes. As reported in this study, *Rhizophora apiculata* and *Rhizopohora mucronata* dominate most planted sites, whereas natural forests host a broader range of mangrove species that create diverse microhabitats (Bosire et al., 2004; Friess et al., 2019). A diverse forest structure influences sediment texture, leaf litter input, and root morphology, all of which are essential for supporting benthic invertebrate diversity.

Moreover, natural mangrove forests exhibit better regeneration, as indicated by higher seedling density. This contributes to ecosystem resilience and habitat complexity, supporting a more stable and diverse faunal community (Lee, 2008). Conversely, the planted mangroves in this study showed limited regeneration and a low Simpson's diversity index for mollusks (0.75–0.90) and crabs (0.48–0.90) compared with values above 1.4 in naturally growing mangroves. This suggests that while planted mangroves may look structurally "dense," they lack the ecological depth to support rich faunal (crabs and mollusks) assemblages.

Environmental heterogeneity also plays a crucial role. Naturally, growing mangroves develop under dynamic hydrological, sedimentary, and salinity gradients, shaping habitat niches that favor diverse species assemblages (Nagelkerken et al., 2008). In contrast, planted mangroves are often established on homogenized substrates, which may lack the microhabitat variability necessary for supporting diverse faunal communities (Cannicci et al., 2018).

In essence, the trade-off between carbon accumulation and biodiversity in planted mangrove forests highlights a fundamental consideration in mangrove restoration and rehabilitation. Restoration projects should prioritize not only carbon targets but also multispecies planting and habitat complexity to foster faunal diversity and broader ecological functioning.

Implications for Mangrove Restoration and Management

The results of this study have important implications for mangrove restoration and coastal ecosystem management in Indonesia and beyond. First, the demonstrated capacity of planted mangroves to store large amounts of carbon, especially AGC, makes them valuable for climate change mitigation initiatives such as blue carbon strategies and carbon offset schemes (Donato et al., 2011; Howard et al., 2017; Suyadi et al., 2020). However, the predominance of high-density but small-diameter trees suggests that the carbon gains are quantitative rather than qualitative and may plateau or decline as the forest structure matures.

Second, the low species richness of mollusks and crabs in planted mangroves signals the need to rethink current restoration practices. Restoration efforts often emphasize survival rates and canopy closure but neglect ecological integrity, including faunal diversity, trophic interactions, and natural recruitment. This study demonstrated that single-species plantations, although efficient for rapid greening and carbon sequestration, are insufficient for restoring the full ecological functions of mangrove ecosystems (Kodikara et al., 2017).

Third, integrating ecological mangrove restoration (EMR) principles is essential. The EMR promotes natural regeneration and emphasizes site-specific planning, including hydrology, soil characteristics, and reference ecosystems (Lewis, 2005). Restoration projects should favor mixed-species planting, guided by local species assemblages and ecological zoning, to create structurally and functionally resilient forests. Incorporating species such as Avicennia, Bruguiera, and Sonneratia can diversify root structure, leaf litter quality, and habitat niches, ultimately supporting richer faunal diversity.

Fourth, monitoring programs should include not only vegetation growth metrics but also biodiversity indicators, especially for benthic fauna and fish nurseries. These organisms are bioindicators of habitat quality and can inform adaptive management approaches. For example, declining crab or mollusk diversity may indicate problems with sediment oxygenation, litter availability, or hydrological connectivity (Cannicci et al., 2008).

Finally, the findings emphasize the need for long-term management and community engagement. Local stakeholders must be empowered with knowledge and tools to support not only planting but also postplanting maintenance, enrichment planting, and biodiversity monitoring. Restoration should be seen not as a one-time intervention but as a process of ecological recovery, requiring continuous adjustment and learning.

In conclusion, while mangrove planting significantly enhances carbon storage, particularly AGC, to achieve carbon emission reduction (NDC target) and economic benefit through blue carbon markets, it should be balanced with efforts to restore biodiversity and ecological function. Future mangrove restoration must exceed survival rates and aim for ecologically complex, resilient, and multifunctional ecosystems.

Conclusion

This study reveals a fundamental trade-off in mangrove restoration outcomes in Indonesia, where planted mangroves (largely established as monocultures) deliver rapid gains in aboveground carbon through high stem density but fail to recover the structural complexity, regeneration capacity, and faunal diversity characteristic of naturally growing forests. Although planted stands store substantially more carbon in the short term, this advantage is dominated by small-diameter trees and may not be sustained as forest structure develops. More importantly, the markedly lower diversity of mollusks and crabs underscores that carbon-focused restoration alone is insufficient to restore mangrove ecosystem function. These findings demonstrate that successful mangrove restoration must be evaluated not only by canopy cover or carbon metrics but also by biodiversity and ecological integrity. Future restoration efforts should therefore prioritize ecologically informed, multispecies approaches and long-term monitoring to ensure that mangrove rehabilitation contributes simultaneously to climate mitigation, biodiversity conservation, and coastal resilience.

Acknowledgments

This research activity is financially supported through RIIM (Riset Inovasi Indonesia Maju) Batch 1 from Indonesia Fund for Education Agency (LPDP) Ministry of Finance of the Republic of Indonesia and National Research and Innovation Agency of Indonesia (BRIN) according to the contract number: B-803/II.7.5/FR/6/2022 and B-1373/III.5/PR.03.08/6/2022. We would also thanks to the heads of the Krida Wana Lestari of forest farmer group, Mr. Wahyono and Mr. Joni Rianto, for their assistance in the research and data collection process.

Disclaimer

None.

References

Alongi, D. M. (2002). Present state and future of the world's mangrove forests. *Environmental Conservation*, 29(3), 331–349. https://doi.org/10.1017/S0376892902000231.

Alongi, D. M. (2008). Mangrove forests: Resilience, protection from tsunamis, and responses to global climate change. *Estuarine, Coastal and Shelf Science*, 76(1), 1–13. https://doi.org/10.1016/j.ecss.2007.08.024.

Alongi, D. M. (2012). Carbon sequestration in mangrove forests. *Carbon Management*, 3(3), 313–322. https://doi.org/10.4155/cmt.12.20.

Barbier, E. B., Hacker, S. D., Kennedy, C., Koch, E. W., Stier, A. C., & Silliman, B. R. (2011). The value of estuarine and coastal ecosystem services. *Ecological Monographs*, 81(2), 169–193. https://doi.org/10.1890/10-1510.1.

Bosire, J. O., Dahdouh-Guebas, F., Kairo, J. G., & Koedam, N. (2004). Colonization of nonplanted mangrove species into restored mangrove stands in Gazi Bay, Kenya. *Aquatic Botany*, 76(4), 267–279.

Cannicci, S., Burrows, D., Fratini, S., Smith III, T. J., Offenberg, J., & Dahdouh-Guebas, F. (2008). Faunal impact on vegetation structure and ecosystem function in mangrove forests: A review. *Aquatic Botany*, 89(2), 186–200.

Donato, D. C., Kauffman, J. B., Murdiyarso, D., Kurnianto, S., Stidham, M., & Kanninen, M. (2011). Mangroves among the most carbon-rich forests in the tropics. *Nature Geoscience*, 4(5), 293–297.

FAO. (2007). *The world's mangroves 1980–2005*. FAO Forestry Paper 153. Rome: Food and Agriculture Organization.

Friess, D. A., Thompson, B. S., Brown, B., Amir, A. A., Cameron, C., Koldewey, H. J., & Sidik, F. (2019). Policy challenges and approaches for the conservation of mangrove forests in Southeast Asia. *Conservation Biology*, 33(3), 525–537.

Giesen, W., Wulffraat, S., Zieren, M., & Scholten, L. (2007). *Mangrove guidebook for Southeast Asia*. FAO and Wetlands International.

Howard, J., Hoyt, S., Isensee, K., Pidgeon, E., & Telszewski, M. (2017). *Coastal blue carbon: methods for assessing carbon stocks and emissions factors in mangroves, tidal salt marshes, and seagrass meadows*. Conservation International.

Keputusan Menteri Negara Lingkungan Hidup (Kepmen LH) Republik Indonesia Nomor 201 Tahun 2004 tentang Kriteria Baku dan Pedoman Penentuan Kerusakan Mangrove.

Kodikara, K. A. S., Mukherjee, N., Jayatissa, L. P., Dahdouh-Guebas, F., & Koedam, N. (2017). Have mangrove restoration projects been successful? An in-depth study in Sri Lanka. *Restoration Ecology*, 25(5), 705–716.

Lee, S. Y. (2008). Mangrove macrobenthos: Assemblages, services, and linkages. *Journal of Sea Research*, 59(1-2), 16–29.

Lewis, R. R. (2005). Ecological engineering for successful management and restoration of mangrove forests. *Ecological Engineering*, 24(4), 403–418.

MoEF (Ministry of Environment and Forestry). (2021). State of Indonesia's Forests 2021. Jakarta: MoEF.

Murdiyarso, D., Donato, D. C., Kauffman, J. B., et al., (2015). The potential of Indonesian mangrove forests for global climate change mitigation. *Nature Climate Change*, 5, 1089–1092. https://doi.org/10.1038/nclimate2734.

Nagelkerken, I., Blaber, S. J. M., Bouillon, S., Green, P., Haywood, M., Kirton, L. G., & Somerfield, P. J. (2008). The habitat function of mangroves for terrestrial and marine fauna: A review. *Aquatic Botany*, 89(2), 155–185.

Rahman, H. E., Rusmana, I., and Ali, M. (2020). Metode Pengukuran dan Model Pendugaan Biomassa Nypa Fruticans di Sungai Tallo, Makassar-Indonesia Measurement Methods and Estimate Models of Nypa Fruticanus Biomass in Tallo River, Makassar-Indonesia. *Jurnal Grouper*, 11(1), 25-30.

Suyadi, Nurdiansah, D., Renyaan, J., Hapsarai, B. W., Rahayu, E. M. D., Sugiharto, A., & Ulumuddin, Y. I. (2023). Better Approaches are needed for successful mangrove restoration and rehabilitation programs. In *AIP Conference Proceedings* (Vol. 2606, No. 1, p. 050001). AIP Publishing LLC. https://doi.org/10.1063/5.0118836.

Suyadi, J. Gao, C. J. Lundquist, L. Schwendenmann. (2020). Aboveground carbon stocks in rapidly expanding mangroves in New Zealand: regional assessment and economic valuation of blue carbon. *Estuaries and Coasts*. doi.org/10.1007/s12237-020-00736-x.

Chapter 6

The Resilience of Coral Species to Environmental Stressors: A Case Study of Kapoposang MPA, Indonesia

Hedi Indra Januar[1,*]
Rinny Rahmania[1]
Suyadi Suyadi[2]
Yulizar Irhami Rahmila[2]
Marwan Setiawan[2]
Parlin Hotmartua Putra Pasaribu[2]
Varenna Faubiany[2]
Mira Yulianti[2]
Izhamil Hidayah[2]
and Zilda Dona Okta Permata[2]

[1]Indonesian Research Centre for Conservation of Marine and Inland Water Resources, Cibinong, West Java, Indonesia
[2]Indonesian Research Centre for Ecology and Ethnobiology, KST Soekarno, Cibinong, Indonesia

Abstract

The Coral Triangle is the global epicenter of marine biodiversity, providing critical ecological and economic value by sustaining tropical fisheries. However, nutrient enrichment from anthropogenic pressures

---

* Corresponding Author's Email: hedi002@brin.go.id

In: Biodiversity
Editors: Mohammad Basyuni and Faujiah Nurhasanah Ritonga
ISBN: 979-8-90134-246-6

and climate change threatens reef resilience, undermining biodiversity and ecosystem sustainability. This study examines coral species resilience and conservation strategies in the Kapoposang Marine Protected Area (MPA), South Sulawesi, as a case study for the Coral Triangle reefs. Seasonal field surveys (west, intermediate, and east monsoon) were used to assess biodiversity, coral cover, and water quality. The results of the present study revealed that elevated phosphate, nitrate, and ammonia concentrations significantly limited coral growth and biodiversity. Areas enriched with nutrients included increased amounts of abiotic substrates, dead coral with algae, and invasive macroalgae. In contrast, reefs with higher dissolved oxygen (DO) contents and stable pH values support greater biodiversity and live coral cover, highlighting the critical role of water quality in maintaining resilience. The patterns of nutrient-driven degradation, underscoring the urgency of targeted interventions, controlling nutrient inputs and safeguarding water quality, are essential for sustaining coral resilience. This case study emphasizes that the persistence of coral reefs in the Coral Triangle depends on effective conservation management, closing enforcement gaps, and establishing continuous monitoring systems. Stronger adaptive management strategies, combined with enhanced enforcement, education, and local capacity-building, are urgently needed. By prioritizing nutrient management and proactive water-quality safeguards, conservation efforts can better protect coral reef ecosystems, ensuring their long-term ecological and economic contributions to the Coral Triangle region.

Keywords: coral reef, Indonesia, marine protected areas, nutrient enrichment, spermonde archipelago, tropical coral triangle, water quality

Introduction

Coral reefs in the Indo-Pacific Coral Triangle represent the global epicenter of marine biodiversity, supporting more than 605 coral species or 75% of global diversity (Álvarez-Filip et al., 2011; Jones et al., 2019). Compared with subtropical reefs, Indo-Pacific systems sustain greater coral cover and resilience, whereas subtropical reefs experience greater algal dominance, which limits coral recruitment and structural integrity (Hoey et al., 2011; McIlroy et al., 2019; Reimer et al., 2022). Ecologically, coral triangle reefs provide vital habitats and foraging grounds for diverse, commercially important species, maintaining biodiversity through complex trophic interactions (Riniwati et al., 2019; Williams et al., 2017). They also provide key services, including coastal protection against erosion and oceanographic forces (Wang et al., 2015). Economically, these reefs sustain more than 120 million people reliant on fisheries and tourism (Marwayana et al., 2021), with an estimated annual value of USD 2.3 billion (Beger et al., 2015).

One of the coral reef areas within the Indo-Pacific Coral Triangle is the Kapoposang Marine Protected Area (MPA). This area is located on the western side of the Coral Triangle region, between Kalimantan and Sulawesi, Indonesia. Coral health in this region is closely linked to the abundance of tropical fishery resources (Riskiani et al., 2019). However, previous studies reported suboptimal coral growth, with coral cover ranging from only 46.73% to 73.08% (Arifin et al., 2020). This indicates that environmental pressures, which may originate from both anthropogenic factors (such as sedimentation, nutrient pollution, and Acanthaster outbreaks, overfishing) and climate change factors (such as bleaching, reduced calcification, and storm damage), threaten the sustainability of coral reef ecosystems (Afandy et al., 2019; Bayraktarov et al., 2019; Beyer et al., 2018; Gibbs et al., 2021; Guan et al., 2020; Denley et al., 2020; Kornder et al., 2018; Novi & Bracco, 2022; Permana et al., 2020; Su et al., 2025; Renggong et al., 2022; Ziegler et al., 2018). Thus, establishing effective marine protected areas and reducing local stressors are critical, as supported by baseline studies and long-term monitoring (Edullantes et al., 2021; Masucci & Reimer, 2019; Schill et al., 2015; Singgalen, 2024). However, no conservation recommendations based on baseline assessments have yet been developed for the Kapoposang MPA.

Therefore, this work aims to provide a baseline assessment of coral biodiversity, cover, and water quality in Kapoposang MPA, evaluate resilience under current pressures, and inform conservation strategies to increase MPA effectiveness. Addressing environmental challenges is critical for sustaining the resilience of coral reefs and fisheries in this biodiversity hotspot. Therefore, conducting a baseline study that includes key indicators of coral reef health, such as coral cover, water quality, and pollution levels, is essential for assessing MPA effectiveness and developing dynamic, time-responsive conservation strategies (Denley et al., 2020; Lam et al., 2017). The findings from this analysis are expected to serve as a model for implementing baseline studies to optimize conservation efforts in other tropical coral reef regions.

Methods

Study Sites

The Kapoposang Marine Protected Area (MPA) spans approximately 500 km^2 and encompasses six main islands: Kapoposang, Papandangan, Gondangbali, Tambakulu, Pamang-Gangang, and Suranti (Figure 1). Data from the Allen Coral Atlas indicate that this marine area contains approximately 34.6 km^2 of benthic habitat, with 42.22% covered by coral/algae, 2.65% covered by seagrass, 11% covered by macroalgae, 16.21% covered by rock, 9.34% covered by rubble, and 18.48% covered by sand (Allen Coral Atlas, 2022). The conservation area at MPA Kapoposang is divided into four zones: the core conservation zone, the fisheries utilization zone, the general utilization zone for recreational activities, and the rehabilitation zone (Ministry of Marine Affairs and Fisheries, 2014). On the basis of these functional designations, three stations were selected on the western side of Kapoposang Island to represent the core conservation zone. Another three stations were selected on the eastern and southern sides of Kapoposang Island as part of the recreational utilization zone. Moreover, three stations were located in the rehabilitation zone near Papandangan Island. The remaining six stations were situated within the fisheries utilization zone, covering the waters around Gondangbali and Tambakulu Islands, where local fishers operate. Monitoring activities were conducted three times a year: in March (during the west monsoon), in June (during the east monsoon), and in September (during the transitional period from the east to the west monsoon).

Water Quality Analysis

Seawater samples were collected from the bottom layer of the water column above the transect line for coral cover analysis (Januar et al., 2019). The analysis (pH, dissolved oxygen (DO), salinity, nitrate, ammonia, and phosphate concentrations) was conducted on site via digital portable devices (Hach DR-890, Hach HQ40d with DO and pH probes, and Eutech Salt +6 Salt-meter). Three replicates of water samples were taken at each station.

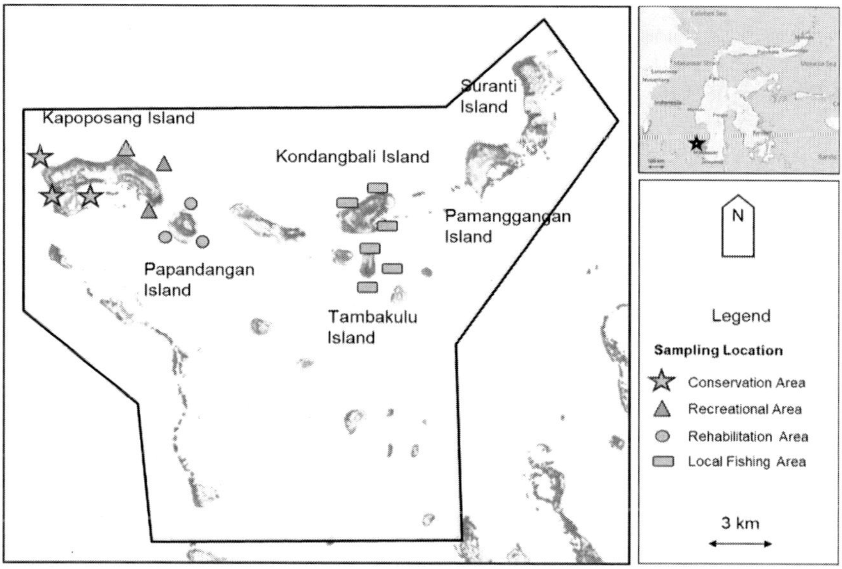

Figure 1. Stations for water quality and coral cover research at conservation, recreational, rehabilitation, and local fishing areas at Kapoposang MPA, South Sulawesi Province, Indonesia.

Coral Cover and Biodiversity Analyses

The coral cover at each site was analyzed at depths of 4–6 m by employing three 30 m long line intercept transects that were parallel and 3 m apart. Underwater photographs were taken via a digital camera every meter on both sides of the transect line with a 0.5 × 0.5 m quadrant frame. Analysis of coral cover was performed with CPCe 3.6 software (Kohler and Gill, 2006). Fifty random points were automatically placed on each photo, and each point was then classified into one of five main groups, i.e., hard coral, other biota (sponge, soft coral, and ascidian), algae, and abiotic (dead coral, rock, and sand).

Data Processing

The Kruskal–Wallis test was employed to assess differences in variables across seasons, islands, and conservation zones. Further multivariate discriminant analysis and canonical correspondence analysis were deployed to analyze the variance and correlation between the coral cover and water variables. All variables were transformed (logarithmic for environmental data and square root or coral cover) to meet the normal distribution data prior to multivariate statistical analyses (Huang et al., 2011; Loayza-Muro et al., 2010). All the statistical analyses were conducted with Past Statistical Software v3.08 (Hammer et al., 2001).

Results

Water Quality

The water quality analysis generally revealed good environmental conditions, particularly high levels of dissolved oxygen (DO), pH and salinity (Figure 2). However, nitrate and phosphate concentrations were found to be relatively high overall, exceeding the marine water quality standards for marine biota established by the Indonesian Ministry of Environment (Ministry of Environment, 2004). Nonsignificant ($P > 0.05$) temporal variation was detected for nutrient concentrations. Moreover, relatively low ($P < 0.05$) nutrient levels were detected around Tambakulu Island. In terms of zoning, the core conservation zone presented the lowest ($P < 0.05$) nutrient concentrations.

Coral Cover and Biodiversity

The data revealed that the rehabilitation zone had the highest ($P < 0.05$) coral cover, averaging nearly 50% of the transect length (Figure 3). The conservation zone had the second-highest coral cover, with an average of approximately 40% across the three temporal monitoring periods conducted at different times. In contrast, the recreational and fishing zones presented low coral cover and were characterized by greater rubble presence than the other zones were. The biodiversity of other benthic groups, such as ascidians, sponges, and soft corals was highest in the conservation and recreational zones. However, algae were also detected at relatively high abundances across all zones and were significantly ($P < 0.05$) greater in the rehabilitation zone. In the fishing, rehabilitation, and recreational zones, the average algal cover exceeded 10% of the transect length.

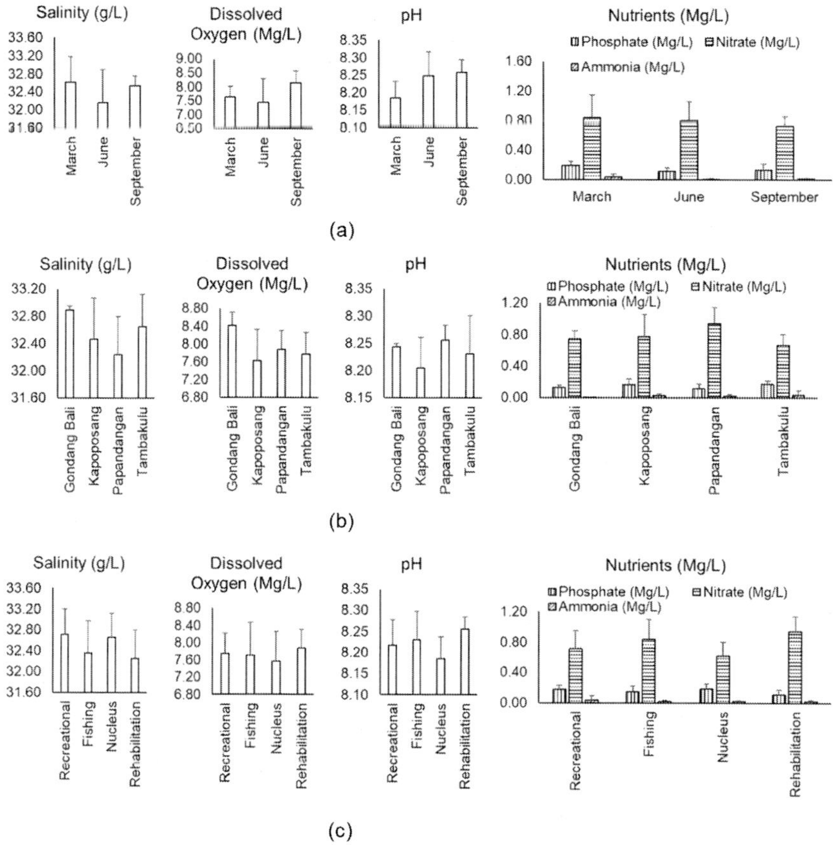

Figure 2. Average water quality (salinity, dissolved oxygen, pH, phosphate, nitrate, and ammonia) in different zones: (a) seasons (March, June, and September), (b) islands (Gondang Bali, Kapoposang, Papandangan, and Tambakulu), and (c) zones (Recreational, Fishing, Conservation, and Rehabilitation) in the Kapoposang Marine Protected Area, Province South Sulawesi, Indonesia.

Moreover, multivariate analysis of water quality and coral cover data revealed the distinct characteristics of each zone within the Kapoposang MPA (Figure 4). The water analysis revealed that, compared with the other three zones, only the rehabilitation zone presented different water characteristics, which were characterized by high nutrient (nitrate) levels and elevated pH. The remaining three zones had overlapping profiles, with the recreational and fishing zones displaying high nutrient concentrations, whereas the conservation zone presented relatively lower nutrient levels. On the other hand, discriminant multivariate analysis of the coral cover patterns revealed key trends: high rubble cover in the fishing zone, high sand cover in the recreational zone, and a combination of hard coral and dead coral with algae in the conservation and rehabilitation zones.

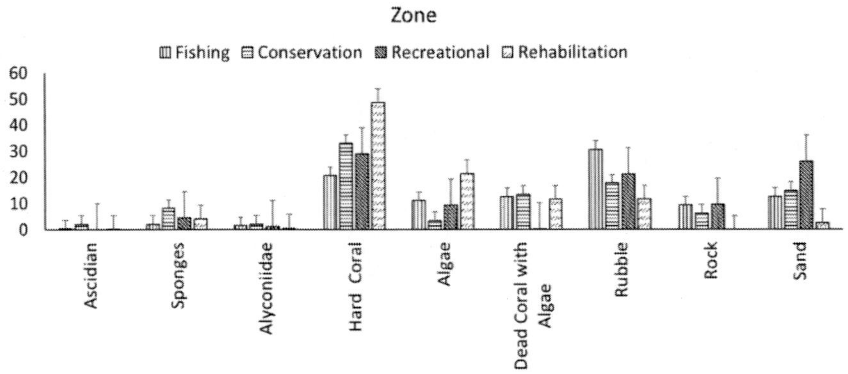

Figure 3. Average coral cover (ascidian, sponges, alyconiidae, hard coral, algae, dead coral with algae, rubble, rock, and sand) at the 3 × 30 m transect line at fishing, conservation, recreational, and rehabilitation zones in the Kapoposang Marine Protected Area, Province South Sulawesi, Indonesia.

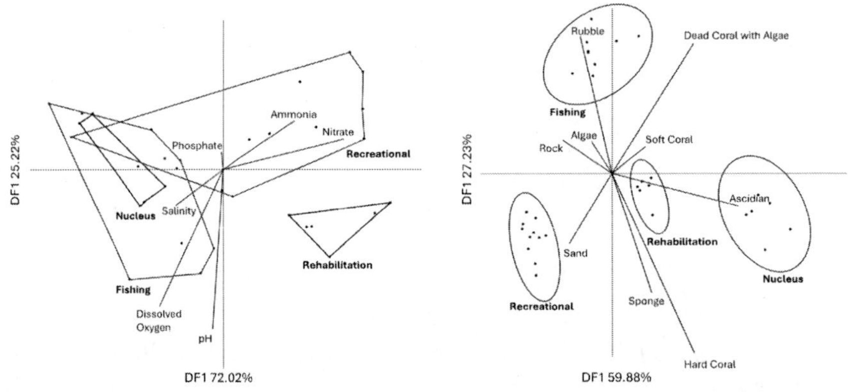

Figure 4. Multivariate discriminant analysis of (a) water quality (pH, DO, salinity, ammonia, nitrate, and phosphate) and (b) coral cover (hard coral, rubble, dead coral with algae, soft coral, sand, rock, and algae) in different zones (fishing, recreational, rehabilitation, and conservation) at Kapoposang MPA, South Sulawesi Province, Indonesia.

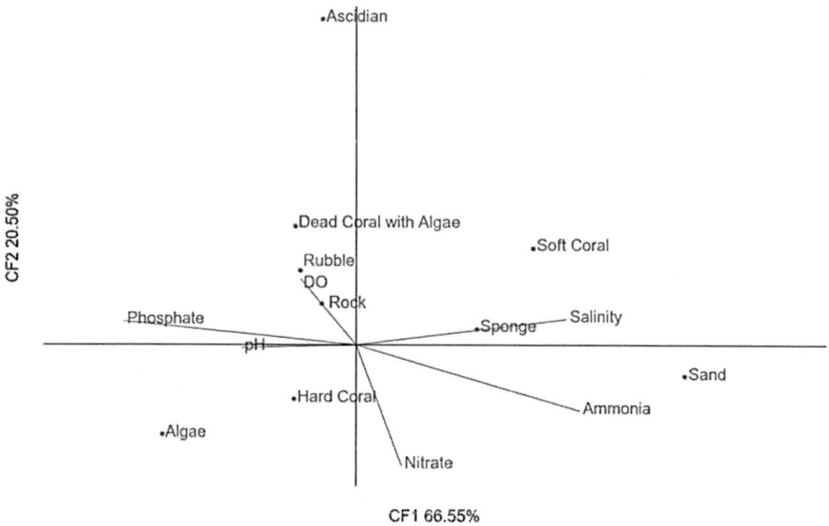

Figure 5. Multivariate canonical correspondence analysis between water quality (pH, DO, salinity, ammonia, nitrate, and phosphate) and coral cover (hard coral, rubble, dead coral with algae, soft coral, sand, rock, and algae) at Kapoposang MPA, South Sulawesi Province, Indonesia.

Further multivariate statistical testing via correspondence analysis revealed relationships between water quality, coral cover, and reef biodiversity (Figure 5). Hard coral was positioned at the center of the dendrogram, indicating that its optimal presence occurred under low-nutrient conditions. Shifts in its composition, particularly with increased phosphate levels, were associated with an increase in algal cover. Elevated levels of dissolved oxygen (DO), which may indicate stronger wave activity, were linked to an increase in rubble and dead coral with algae. Moreover, higher concentrations of other nutrient variables, such as ammonia and nitrate, were more closely associated with the presence of other benthic organisms, such as sponges and soft corals.

Discussion

The dissolved oxygen (DO), pH, and salinity levels in Kapoposang MPA fall within the normal range for a healthy aquatic environment. Normal salinity levels suggest typical small island marine conditions without freshwater intrusion. Moreover, the stability of DO and pH levels indicates that photosynthesis and respiration processes within the water system function well (Breitburg et al., 2018). However, the elevated concentrations of phosphate and nitrate point to an imbalance in the concentrations of the environmental variables. Phosphate ions in coastal waters may originate from the weathering of rocks and calcareous sediments into the water body (Iijima et al., 2021). Visually, the northern side of Kapoposang Island, which consists mostly of sandy flats, is suspected to be a source of such phosphorus ions. Oceanic upwelling may also transport nutrient-rich waters from deeper layers of the ocean to the surface, contributing to elevated nitrate concentrations (Kobari et al., 2020; McGillicuddy et al., 2003). In addition, increased nutrient concentrations are often driven by anthropogenic factors such as domestic wastewater, agricultural runoff, and aquaculture practices (González-Rivas et al., 2020). Inadequate waste management infrastructure on small islands further contributes to excessive nutrient loads in surrounding marine environments (Eldridge et al., 2016).

Moreover, the benthic cover in the Kapoposang MPA was diverse. Relatively good coral cover was observed in the conservation and rehabilitation areas. Moreover, sponges, Alcyoniidae soft corals, sponges, and ascidians, which are vividly colored and found along steep slopes, represent key attractions for underwater marine tourism in the recreational zone. However, from a quantitative perspective, the coral cover reached only 50%, accompanied by the presence of dead coral rubble and algae, each accounting for up to 30% and 20%, respectively, indicating moderate ecological stress within the reef ecosystem. The presence of dead coral overgrown with algae, which reached up to 15% across various zones, highlights spatial competition between coral organisms and algae, driven by environmental conditions and anthropogenic pressures (Beeden et al., 2015). Additionally, the high rubble cover suggested ongoing disturbance impacts on the coral reef ecosystem, typically resulting from hydrological events or anthropogenic influences such as sedimentation and overfishing (Halford & Caley, 2009).

The transition from hard coral-dominated reefs to systems dominated by soft corals, sponges, and algae represents a major ecological shift, which is driven primarily by nutrient dynamics and water quality. Compared with other regions, this issue is a significant stressor for coral reef ecosystems in tropical areas, particularly those affected by anthropogenic activities such as coastal development, agriculture, and urbanization. Coral reefs in tropical coral triangle regions have already experienced considerable losses in marine biodiversity, with declines of up to 60% reported in Indonesian waters (Guan et al., 2020). Research on the water quality and occurrence of *Acropora spp.*

in Maluku waters in the Coral Triangle region revealed that chlorophyll, an indicator of high nutrient levels in the environment, has the highest correlation with the optimal coral habitat (Januar et al., 2023). Terrestrial runoff, exacerbated by inadequate waste management, contributes significantly to high pollutant loads in coastal waters (Januar et al., 2017).

The impacts of increased nutrient levels on coral reefs are complex; hard corals are highly specialized to thrive in oligotrophic waters. Nutrient enrichment or eutrophic conditions can lead to increased coral mortality and reduced reproductive success (Marangoni et al., 2020). Moreover, elevated nutrient concentrations in the water column can stimulate algal growth, which rapidly competes with corals for space and resources, often resulting in a shift toward macroalgal dominance. Such community shifts from hard coral to algal-dominated benthic zones have been widely documented across various coral reef ecosystems (Liu et al., 2015; Spring & Williams, 2023). In addition to algae, soft corals, sponges, and corallimorphs exhibit greater physiological resilience and adaptability under nutrient-rich conditions, further inhibiting the growth and survival of hard corals. This ultimately undermines the biodiversity and structural integrity of coral reef ecosystems (Ali et al., 2013; Alvin et al., 2023; Baum et al., 2016; Holcomb et al., 2010).

Therefore, effective management strategies are urgently needed to reduce nutrient inputs and promote environmental conditions conducive to the recovery and survival of hard corals in these critically important marine ecosystems (Allgeier et al., 2020; Baum et al., 2016). The increasing nutrient load in coastal ecosystems is largely attributable to unmanaged domestic wastewater, agricultural runoff, and industrial discharge, all of which have detrimental effects on coral health and resilience (Kroon et al., 2014; Shaver et al., 2022). Consequently, a crucial first step in addressing these challenges involves the establishment of effective wastewater treatment systems. Restoring water quality to suitable levels will create more favorable conditions for the survival of hard corals, which are highly sensitive to eutrophication. Numerous studies have demonstrated that improving water quality can increase coral resilience and support biodiversity within these ecosystems (Lesser, 2021; Zaneveld et al., 2016). Once nutrient levels are under control, targeted coral ecosystem restoration becomes essential. In areas where community shifts toward algal or soft coral dominance have occurred, transplanting ecologically tolerant hard coral genera such as Porites or Montipora has proven effective. The use of artificial substrates as growth surfaces, alongside coral gardening techniques, can increase coral fragment survival during restoration efforts (Hein et al., 2021; Vardi et al., 2021).

Community engagement in conservation activities has been shown to significantly increase the success of coral reef restoration. Active participation in water quality monitoring and coral cover assessments through citizen science initiatives can empower local communities and improve their adaptive capacity in reef stewardship (Hein et al., 2017; Uribe-Castañeda et al., 2024). Furthermore, sustained environmental education and the development of alternative livelihoods, such as reef-based ecotourism, can provide incentives for communities to support ecosystem sustainability while strengthening their economic stability (Wu, 2024). A combined approach involving nutrient input control, coral ecosystem rehabilitation focused on ecologically resilient species, and the empowerment and active participation of local communities is expected to restore the ecological function and biodiversity of coral reefs while also ensuring long-term livelihood security for coastal populations. This integrated strategy represents a holistic management framework to address the mounting pressures facing coral reef ecosystems in an increasingly dynamic environmental landscape (Calle-Triviño et al., 2025; Caruso et al., 2021).

Study Limitations

This study is limited by only three sampling events within a year, which may not adequately capture temporal variability in coral reef conditions. Seasonal dynamics such as monsoons, tidal cycles, and episodic nutrient runoff likely influence coral health but have not been fully characterized. Spatial coverage was also restricted, potentially overlooking zones with greater degradation or resilience. Future research should extend over multiple years with higher sampling frequencies (e.g., quarterly or monthly) and incorporate continuous monitoring via automated loggers for temperature, turbidity, and nutrients. Broader spatial coverage across reef zones would better reflect ecosystem heterogeneity. Long-term studies linking reef health and water quality to socioeconomic drivers, including fishing and land-use change, are also essential. Finally, molecular approaches such as coral microbiome profiling and identification of nutrient-reducing bacteria can provide early stress indicators and bioindicators to inform adaptive, ecosystem-based conservation strategies.

Conclusion

The study at Kapoposang MPA underscores how nutrient enrichment limits coral growth despite supportive conditions of dissolved oxygen (DO) and pH. Elevated phosphate and nitrate concentrations promote algal dominance and soft coral expansion, leading to ecological shifts away from hard corals, which are essential for reef structure and biodiversity. These findings highlight the urgency of managing nutrient inputs, as eutrophication reduces coral survival and accelerates reef degradation. Effective conservation requires integrated strategies. Priority should be given to improving wastewater treatment, regulating agricultural runoff, and addressing land-based pollution to restore water quality. Restoration should focus on resilient hard coral genera, such as Porites and Montipora, supported by artificial substrates and coral gardening. Moreover, community involvement in monitoring, restoring, and sustainable ecotourism is vital for long-term stewardship. Combining nutrient control, ecological restoration, and community engagement provides a holistic framework to sustain coral resilience and biodiversity at the Coral Triangle.

Disclaimer

None.

References

Álvarez-Filip, L., Dulvy, N. K., Côté, I. M., Watkinson, A. R., & Gill, J. A. (2011). Coral identity underpins architectural complexity on Caribbean reefs. *Ecological Applications*, 21(6), 2223-2231.

Afandy, Z., Damar, A., & Agus, S. (2019). Coral recruitment spatial distribution in its relation with coral cover and herbivorous fish abundance within conservation area. *Ilmu Kelautan Indonesian Journal of Marine Sciences*, 24(2). 91-98.

Ali, A., Ormond, R., Leujak, W., & Siddiqui, P. J. A. (2013). Distribution, diversity and abundance of coral communities in the coastal waters of Pakistan. *Journal of the Marine Biological Association of the United Kingdom*, 94(1), 75-84.

Allen Coral Atlas. (2022). Imagery, maps and monitoring of the world's tropical coral reefs. doi.org/10.5281/zenodo.3833242.

Allgeier, J. E., Andskog, M. A., Hensel, E., Appaldo, R., Layman, C. A., & Kemp, D. W. (2020). Rewiring coral: anthropogenic nutrients shift diverse coral–symbiont nutrient and carbon interactions toward symbiotic algal dominance. *Global Change Biology*, 26(10), 5588-5601.

Alvin, A., Aju, K. R., Sreenath, K. R., Pradeep, M. A., Nisha, E. A., & Joshi, K. K. (2023). Record of a central indo-pacific corallimorph in the lakshadweep atolls and its competition with the hard corals. *Indian Journal of Fisheries*, 70(1), 50-56.

Arifin, T., Rahmania, R., Gunawan, D. P., Setyawidati, N. A., Gusmawati, N., & Ramadhan, M. (2020). Changes of coral reefs condition in the core zones of Kapoposang MPA, Makassar straits. *IOP Conference Series Earth and Environmental Science*, 486(1), 012019.

Baum, G., Januar, I., Ferse, S. C. A., Wild, C., & Kunzmann, A. (2016). Abundance and physiology of dominant soft corals linked to water quality in Jakarta bay, Indonesia. *PeerJ*, 4, e2625.

Bayraktarov, E., Stewart-Sinclair, P. J., Brisbane, S., Boström-Einarsson, L., Saunders, M. I., Lovelock, C. E., ... & Wilson, K. A. (2019). Motivations, success, and cost of coral reef restoration. *Restoration Ecology*, 275: 981-991.

Beeden, R., Maynard, J., Puotinen, M., Marshall, P., Dryden, J., Goldberg, J., & Williams, G. (2015). Impacts and recovery from severe tropical cyclone Yasi on the Great Barrier Reef. *PloS one*, 10(4), e0121272.

Beger, M., McGowan, J., Treml, E. A., Green, A. L., White, A. T., Wolff, N. H., ... & Possingham, H. P. (2015). Integrating regional conservation priorities for multiple objectives into national policy. *Nature Communications*, 6(1), p8208.

Beyer, H. L., Kennedy, E. V., Beger, M., Chen, C. A., Cinner, J. E., Darling, E. S., ... & Hoegh-Guldberg, O. (2018). Risk-sensitive planning for conserving coral reefs under rapid climate change. *Conservation Letters*, 11(6), e12587.

Breitburg, D., Levin, L. A., Oschlies, A., Grégoire, M., Chavez, F. P., Conley, D. J., ... & Zhang, J. (2018). Declining oxygen in the global ocean and coastal waters. *Science*, 359(6371), eaam7240.

Calle-Triviño, J., Rojas-Cano, D., Niño-Torres, L. A., Colín-García, N., Hernández-Landa, R. C., Blanco-Pimentel, M., ... & Rioja-Nieto, R. (2025). Rebuilding Coral Reefs: How Tourism Can Be a Driver Behind Solutions in a Changing Ocean. *Diversity*, 17(4), 268.

Denley, D., Metaxas, A., & Scheibling, R. (2020). Subregional variation in cover and diversity of hard coral (Scleractinia) in the Western Province, Solomon Islands following an unprecedented global bleaching event. *Plos one*, 15(11), e0242153.

Edullantes, B., Maglangit, F., Ortiz, A. M., Casibo, J. M. R., Vicentuan, L. L. C., & Bensig, E. O. (2021). A benchmark survey of plankton, fish and benthic composition in Poblacion and Kadurong Reefs in Liloan, Cebu, Philippines. *Biodiversity Data Journal*, 9, e72537.

Eldridge, D. J., Poore, A. G. B., Ruiz-Colmenero, M., Letnic, M., & Soliveres, S. (2016). Ecosystem structure, function, and composition in rangelands are negatively affected by livestock grazing. *Ecological Applications*, 26(4), 1273-1283.

Gibbs, D., West, J., & Bradley, P. (2021). Incorporating adaptation and resilience into an integrated watershed and coral reef management plan. *Plos One*, 16(6), e0253343.

González-Rivas, D. A., Tapia-Silva, F. O., Bustillos-Guzmán, J. J., Revollo-Fernández, D. A., Beltrán-Morales, L. F., Lluch-Cota, D. B., & Ortega-Rubio, A. (2020). Estimating nitrogen runoff from agriculture to coastal zones by a rapid GIS and remote sensing-based method for a case study from the irrigation district Río Mayo, Gulf of California, Mexico. *Frontiers in Marine Science*, 7, 316.

Guan, Y., Hohn, S., Wild, C., & Merico, A. (2020). Vulnerability of global coral reef habitat suitability to ocean warming, acidification and eutrophication. *Global Change Biology*, 26(10), 5646-5660.

Halford, A. R. & Caley, M. J. (2009). Toward an understanding of resilience in isolated coral reefs. *Global Change Biology*, 15(12), 3031-3045.

Hammer O., Harper D. A. T., & Ryan P. D. (2001). Past: Paleontological Statistics Software package for education and data analysis. *Paleontological Electronic*, 4(1), 9pp.

Hein, M. Y., Willis, B. L., Beeden, R., & Birtles, A. (2017). The need for broader ecological and socioeconomic tools to evaluate the effectiveness of coral restoration programs. *Restoration Ecology*, 25(6), 873-883.

Hein, M. Y., Vardi, T., Shaver, E. C., Pioch, S., Boström-Einarsson, L., Ahmed, M., Grimsditch, G. & McLeod, I. M. (2021). Perspectives on the use of coral reef restoration as a strategy to support and improve reef ecosystem services. *Frontiers in Marine Science*, 8, 618303.

Hoey, A. S., Pratchett, M. S., & Cvitanovic, C. (2011). High macroalgal cover and low coral recruitment undermines the potential resilience of the world's southernmost coral reef assemblages. *Plos One*, 6(10), e25824.

Holcomb, M., McCorkle, D. C., & Cohen, A. L. (2010). Long-term effects of nutrient and CO_2 enrichment on the temperate coral *Astrangia poculata* (ellis and solander, 1786). *Journal of Experimental Marine Biology and Ecology*, 386 (1-2), 27-33.

Huang, Y. C., Hsieh, H. J., Huang, S. C., Meng, P. J., Chen, Y. S., Keshavmurthy, S., ... & Chen, C. A. (2011). Nutrient enrichment caused by marine cage culture and its influence on subtropical coral communities in turbid waters. *Marine Ecology Progress Series*, 423, 83–93.

Iijima, M., Yasumoto, J., Iguchi, A., Koiso, K., Ushigome, S., Nakajima, N., ... & Yasumoto, K. (2021). Phosphate bound to calcareous sediments hampers skeletal development of juvenile coral. *Royal Society Open Science*, 8(3), 201214.

Januar, H. I., Zamani, N. P., Soedarma, D., Chasanah, E., & Wright, A. D. (2017). Tropical coral reef coral patterns in Indonesian shallow water areas close to underwater volcanic vents at Minahasa Seashore, and Mahengetang and Gunung Api Islands. *Marine ecology*, 38(2), e12415.

Januar, H. I., Pratitis, A., & Susilowati, R. (2019). Statistical multimetric-based eutrophication index: Case of study at Batam Marine Reserve Park. *Ilmu Kelautan: Indonesian Journal of Marine Sciences*, 24(4), 164-170.

Januar, H. I., Hidayah, I., Humaida, N., Iswani, S., & Hidayat, A. (2023). Habitat suitability modeling of Acropora spp. distribution in Coral Triangle area of Maluku Waters, Indonesia under influence of future climate change and coastal pollution. *Agriculture and Natural Resources*, 57(5), 869-876.

Jones, L. A., Mannion, P. D., Farnsworth, A., Valdes, P. J., Kelland, S., & Allison, P. A. (2019). Coupling of paleontological and neontological reef coral data improves forecasts of biodiversity responses under global climatic change. *Royal Society Open Science*, 6(4), 182111.

Kobari, T., Honma, T., Hasegawa, D., Yoshie, N., Tsutsumi, E., Matsuno, T., ... & Nakamura, H. (2020). Phytoplankton growth and consumption by microzooplankton stimulated by turbulent nitrate flux suggest rapid trophic transfer in the oligotrophic kuroshio. *Biogeosciences*, 17(9), 2441-2452.

Kohler K. E., Gill S. M. (2006). Coral Point Count with Excel extensions (CPCe) : A visual basic program for the determination of coral and substrate coverage using random point coral methodology. *Computational Geoscience*, 32, 1259-1269.

Kornder, N., Riegl, B., & Figueiredo, J. (2018). Thresholds and drivers of coral calcification responses to climate change. *Global Change Biology*, 24(11), 5084-5095.

Kroon, F. J., Schaffelke, B., & Bartley, R. (2014). Informing policy to protect coastal coral reefs: insight from a global review of reducing agricultural pollution to coastal ecosystems. *Marine Pollution Bulletin*, 85(1), 33-41.

Lam, V., Doropoulos, C., & Mumby, P. (2017). The influence of resilience-based management on coral reef monitoring: a systematic review. *Plos One*, 12(2), e0172064.

Lesser, M. P. (2021). Eutrophication on coral reefs: what is the evidence for phase shifts, nutrient limitation and coral bleaching. *BioScience*, 71(12), 1216-1233.

Liu, P. J., Hsin, M. C., Huang, Y. H., Fan, T. Y., Meng, P. J., Lu, C. C. & Lin, H. J. (2015). Nutrient enrichment coupled with sedimentation favors sea anemones over corals. *Plos One*, 10(4): e0125175.

Loayza-Muro R. A., Elías-Letts R., Marticorena-Ruíz J. K., Palomino E. J., Duivenvoorden J. F., Kraak M. H. S., & Admiraal W. (2010). Metal-induced shifts in benthic macroinvertebrate community composition in Andean high altitude streams. *Environmental Toxicology and Chemistry*, 29(12), 2761–2768.

Fernandes de Barros Marangoni, L., Ferrier-Pagès, C., Rottier, C., Bianchini, A., & Grover, R. (2020). Unraveling the different causes of nitrate and ammonium effects on coral bleaching. *Scientific Reports*, 10(1), 11975.

Marwayana, O. N., Gold, Z., Meyer, C. P., & Barber, P. H. (2022). Environmental DNA in a global biodiversity hotspot: Lessons from coral reef fish diversity across the Indonesian archipelago. *Environmental DNA*, 4(1), 222-238.

Masucci, G. & Reimer, J. (2019). Expanding walls and shrinking beaches: loss of natural coastline in Okinawa Island, Japan. *Peerj*, 7, e7520.

McIlroy, S. E., Thompson, P., Yuan, F. L., Bonebrake, T. C., & Baker, D. M. (2019). Subtropical thermal variation supports persistence of corals but limits productivity of coral reefs. *Proceedings of the Royal Society B: Biological Sciences*, 286(1907), 20190882.

McGillicuddy, D. J., Anderson, L., Doney, S. C., & Maltrud, M. (2003). Eddy-driven sources and sinks of nutrients in the upper ocean: results from a 0.1° resolution model of the north atlantic. *Global Biogeochemical Cycles*, 17(2), 1035.

Minister of Marine Affairs and Fisheries. (2014). Decree of the Minister of Marine Affairs and Fisheries of the Republic of Indonesia No. 59/KEPMEN-KP/2014 concerning the Management and Zoning Plan for the Kapoposang Islands and Surrounding Seas Marine Tourism Park in South Sulawesi Province for the Years 2014–2034. *Ministry of Marine Affairs and Fisheries*, Jakarta - Indonesia.

Minister of Environment of Indonesia. (2004). Decree of the State Minister for the Environment No. 51 of 2004 concerning Seawater Quality Standards. *Deputy Minister for Environment: Division of Environmental Policy and Institutional Affairs*, Jakarta - Indonesia.

Novi, L. & Bracco, A. (2022). Machine learning prediction of connectivity, biodiversity and resilience in the Coral Triangle. *Communications Biology*, 5(1), 1359.

Permana, R., Akbarsyah, N., Putra, P., & Andhikawati, A. (2020). Analysis condition of coral reef covering in pramuka island waters, seribu islands using line intercept transect (lit) method. *Journal of Biological Research and Its Application*, 2(2), 77-81.

Reimer, J. D., Albelda, R. L., Biondi, P., Hardianto, E., Huang, S., Masucci, G. D., ... & Zhu, Y. (2022). Literature review of coral reef restoration in and around the coral triangle from the viewpoint of marine biodiversity. *Arquivos De Ciências Do Mar*, 55(Especial), 413-431.

Renggong, R., Hamid, A., & Yulia, Y. (2022). Investigating law enforcement for coral reef conservation of the spermonde archipelago, indonesia. *Asian Journal of Conservation Biology*, 11(1), 3-11.

Riniwati, H., Harahab, N., & Abidin, Z. (2019). A vulnerability analysis of coral reefs in coastal ecotourism areas for conservation management. *Diversity*, 11(7), 107.

Riskiani, I., Budimawan, B., & Bahar, A. (2019). The analysis of coral reef fishes abundance based on coral reef condition in marine tourism park of the kapoposang islands, south sulawesi, indonesia. *International Journal of Environment Agriculture and Biotechnology*, 4(4), 1012-1017.

Schill, S., Raber, G., Roberts, J., Treml, E., Brenner, J., & Halpin, P. (2015). No reef is an island: integrating coral reef connectivity data into the design of regional-scale marine protected area networks. *Plos One*, 10(12), e0144199.

Shaver, E. C., McLeod, E., Hein, M. Y., Palumbi, S. R., Quigley, K., Vardi, T., ... & Wachenfeld, D. (2022). A roadmap to integrating resilience into the practice of coral reef restoration. *Global Change Biology*, 28(16), 4751-4764.

Singgalen, Y. (2024). Coral database and monitoring system design for ecological sustainability. *Journal of Information Systems and Informatics*, 6(1), 270-289.

Spring, D. L., & Williams, G. J. (2023). Influence of upwelling on coral reef benthic communities: a systematic review and meta-analysis. *Proceedings of the Royal Society B*, 290(1995), 20230023.

Su, S., Liu, J., Chen, B., Wang, W., Xiao, J., Li, Y., ... & Zhang, J. (2025). Distribution Shifts of Acanthaster solaris Under Climate Change and the Impact on Coral Reef Habitats. *Animals*, 15(6), 858.

Uribe-Castañeda, N., Morales, N., Loizzo, J., Patterson, J. T., & Main, M. B. (2024). Community engagement in coral reef restoration in the seaflower MPA, Colombia. *Restoration Ecology*, 32(8), e14268.

Vardi, T., Hoot, W. C., Levy, J., Shaver, E. C., Winters, R. S., Banaszak, A. T., … & Montoya-Maya, P. (2021). Six priorities to advance the science and practice of coral reef restoration worldwide. *Restoration Ecology*, 29(8), e13498.

Wang, M., Ahmadia, G. N., Chollett, I., Huang, C., Fox, H., Wijonarno, A., … & Madden, M. (2015). Delineating biophysical environments of the sunda banda seascape,

Indonesia. *International Journal of Environmental Research and Public Health*, 12(2), 1069-1082.

Williams, S. L., Ambo-Rappe, R., Sur, C., Abbott, J. M., & Limbong, S. R. (2017). Species richness accelerates marine ecosystem restoration in the coral triangle. *Proceedings of the National Academy of Sciences*, 114(45), 11986-11991.

Zaneveld, J. R., Burkepile, D. E., Shantz, A. A., Pritchard, C. E., McMinds, R., Payet, J. P., ... & Thurber, R. V. (2016). Overfishing and nutrient pollution interact with temperature to disrupt coral reefs down to microbial scales. *Nature communications*, 7(1), 11833.

Ziegler, M., Quéré, G., Ghiglione, J., Iwankow, G., Barbe, V., Boissin, É., … & Voolstra, C. (2018). Status of coral reefs of upolu (independent state of samoa) in the south west pacific and recommendations to promote resilience and recovery of coastal ecosystems. *Marine Pollution Bulletin*, 129(1), 392-398.

Chapter 7

Localized Conservation in the Islamic World: Faith, Governance, and Resilience

Roslina Ismail*
Faculty of Business, Economics and Social Development,
Universiti Malaysia Terengganu, Terengganu, Malaysia

Abstract

Despite global frameworks and commitments, biodiversity continues to decline at an unprecedented rate, highlighting the shortcomings of mainstream conservation approaches. This study examines Malaysia's Tasik Kenyir Geopark as a case study of the tensions between economic development and biodiversity protection. It also evaluates wider conservation challenges across Muslim-majority member states (MMMS) of the Organization of Islamic Cooperation (OIC), situating them within global biodiversity patterns and highlighting gaps in current strategies. Employing a qualitative design, the research draws upon secondary sources, including administrative records, scholarly literature, and policy documents, complemented by in-depth interviews with five key informants. The data from the interviews were analyzed via holistic coding and triangulated with secondary sources to identify emergent themes related to governance, ecosystem resilience, and sociocultural drivers of conservation. The findings revealed that, prior to the designation of the Tasik Kenyir Geopark, biodiversity management was shaped primarily by a top-down governance model. More broadly, global conservation frameworks inadequately reflect postcolonial and global Muslim community contexts; in the MMMS, conservation is constrained by geostrategic complexities, postcolonial conservation legacies,

* Corresponding Author's Email: roslina@umt.edu.my

In: Biodiversity
Editors: Mohammad Basyuni and Faujiah Nurhasanah Ritonga
ISBN: 979-8-90134-246-6

ontological and epistemological gaps, and the marginalization of indigenous knowledge. The study advances a pluralistic framework that integrates Islamic spiritual values into policy solutions and state actions in response to global biodiversity loss. This study concludes by recommending conservation practices that balance universal standards with local realities to increase the long-term resilience of ecosystems in the MMMS and other OIC member states and beyond.

Keywords: biodiversity, conservation, OIC, faith, governance

Introduction

Despite the establishment of global frameworks and commitments to address the biodiversity crisis, global biodiversity continues to decline at an unprecedented rate, driven primarily by habitat loss, ecosystem degradation, and uneven conservation outcomes (Abdala--Roberts et al., 2025; Yuan et al., 2024). This trend underscores the inadequacy of current conservation strategies and implementation (Leclère et al., 2020). Recognizing these limitations, Diaz et al., (2019) argue that conservation measures require profound transformation across economic, social, and political domains. Such a claim directly critiques mainstream conservation approaches, which remain largely rooted in Western epistemological frameworks. These frameworks have shaped not only global environmental scholarship but also policy discourse and implementation across the Muslim–majority member states (MMMS) of the Organization of Islamic Cooperation (OIC).

In the Islamic world, Western paradigms continue to dominate both discourse and practice. Murad (2023), for example, contends that contemporary environmental discourse and practice in the MMMS remain heavily influenced by Western models, despite the richness of Islamic ethical resources and institutional conservation traditions. According to Blankinship et al., 2025) references to biodiversity conservation in the Qur'an, sustainable resource management can positively impact Muslim communities locally, nationally, and globally. Although scholars have recognized the abundance of Islamic-based ethical principles and conservation institutions, these resources remain largely underexplored and underutilized in the policies of many OIC member states (Foltz et al., 2003). The persistence of "fortress conservation" models rooted in colonial legacies continues to shape conservation policy, particularly in the MMMS context, where such frameworks were inherited.

However, reliance on these Western-oriented paradigms has not demonstrably improved biodiversity outcomes, even in economically advanced OIC member states (Ismail et al., 2024). Consequently, prevailing conservation frameworks in the MMMS remain anchored in Western-centric paradigms, systematically marginalizing postcolonial, faith-informed, and culturally embedded approaches.

In the Islamic world, Western paradigms continue to dominate discourse and practice. For example, Murad (2023) noted that environmental policies and practices in the MMMS remain heavily influenced by Western models, despite the richness of Islamic ethical traditions and indigenous conservation institutions. While scholars have recognized these resources, they remain largely underexplored and underutilized in OIC policy frameworks (Foltz et al., 2003). The persistence of "fortress conservation" models deeply rooted in colonial legacies further illustrates how inherited Western paradigms continue to shape conservation policy. However, reliance on these approaches has not led to demonstrably better biodiversity outcomes, even in more economically advanced OIC member states (Ismail et al., 2024). Consequently, conservation in the MMMS remains dominated by Western-centric paradigms, systematically marginalizing postcolonial, faith-informed, and culturally embedded approaches.

Given the escalating global environmental crisis, the Islamic world stands at a critical juncture. Global agreements such as the Convention on Biological Diversity (CBD), alongside regional and national conservation initiatives, often remain insufficient to address the root causes of biodiversity loss. In the MMMS context, comparative analysis of Western-based and Islamic conservation approaches remains limited. Kula (2001) highlights the gap between Islamic environmental discourse and policy implementation, whereas Bagader et al., (2010) revisit traditional Hima systems as indigenous models of resource management. An-Naim (2008) explores the tensions between Shariah law, human rights, and environmental justice. More recently, Ismail et al., (2024) showed that low-income OIC member states often outperform wealthier counterparts in conservation outcomes not because of institutional strength but because of reduced development pressures and strategic engagement with global conservation partnerships. Similarly, Hussain and Olawuyi (2025) emphasized the potential of Islamic finance and Hima systems as instruments for sustainable resource management in the MENA region. Thus, faith, resilience, and governance domains, which are often treated separately in the political and social sciences, hold both urgent relevance and the potential for integrative solutions. Integrating these domains can provide a more holistic framework for conservation. A localized approach rooted in Islamic spiritual values offers an alternative to mainstream paradigms while fostering ecological resilience. This concept forms the central argument and intellectual foundation of this chapter.

Taken together, the existing studies highlight a significant research gap: the systematic under exploration of Islamic faith-based, culturally embedded, and postcolonial approaches to biodiversity conservation. This chapter addresses this gap by foregrounding the concept of localized conservation in the Islamic world, with a particular focus on the interplay of faith, governance, and ecological resilience across MMMS. It highlights the underexplored role of the Islamic perspective in shaping biodiversity conservation strategies, thereby contributing to more nuanced strategies and grounded in standing of conservation in Muslim contexts. To illustrate these dynamics, this chapter examines Malaysia's Tasik Kenyir Geopark as a case study of the tensions between economic developments and biodiversity conservation. It further explains broader conservation challenges across MMMS, situating them within global biodiversity patterns and exposing persistent weaknesses in current strategies. On the basis of these insights, this chapter proposes a pluralistic framework that integrates Islamic spiritual values into state policies and conservation actions.

The chapter contributes in three key ways. First, it proposes a pluralistic framework that incorporates Islamic ethical and spiritual values, offering an alternative to dominant Western-centric paradigms. Second, it contributes to postcolonial scholarship by interrogating the persistence of colonial legacies in shaping conservation strategies across the MMMS. Third, it enriches interdisciplinary debates by demonstrating the potential synergy of faith, governance, and ecological resilience as the foundation for holistic, localized conservation. By doing so, it repositions Islamic conservation as a central resource for addressing the global biodiversity crisis rather than a marginal discourse.

Before the chapter's objectives are examined, the following questions must be addressed: What is meant by the "Islamic world"? Is it homogeneous or heterogeneous? What does "conservation" signify in this context? What does "localized" imply? How should the phrase "localized conservation in the Islamic world" be interpreted? Regarding the definition of the "Islamic world," this chapter adopts a practical approach: it refers to countries where Muslims form the majority population and full membership of the OIC, acknowledging that some territories may lack full international recognition. This framework includes 51 MMMS of 57 OIC member states, encompassing diverse geographical and socioeconomic realities. This chapter uses the terms 'Islamic world' and 'Muslim-majority states of the OIC (MMMS)' interchangeably.

With respect to the question of homogeneity, the Islamic world cannot be classified as entirely uniform or wholly fragmented. Instead, it exhibits structural complexity, combining shared historical experiences and spiritual values, producing a dual-natured identity simultaneously homogeneous and heterogeneous. This duality is evident in political identity and sociohistorical evolution. Two features, in particular, unite many OIC member states: a shared history of colonization by Western powers, shaping administrative and developmental frameworks, and spiritual solidarity through Islamic values, which continues to guide social, political, and cultural discourse. With respect to the question of homogeneity, the Islamic world or the MMMS cannot be classified as entirely uniform or wholly fragmented. Instead, it exhibits structural complexity, combining shared historical experiences and spiritual values, producing a dual-natured identity that is simultaneously homogeneous and heterogeneous. This duality is evident in political identity and sociohistorical evolution. Two features, in particular, unite many OIC member states: a shared history of colonization by Western powers, shaping administrative and developmental frameworks, and spiritual solidarity through Islamic values, which continues to guide social, political and cultural discourse. For the former, scholars widely agree that Western colonialism reshaped political, educational, and economic institutions, leaving lasting imprints in the Islamic world. Ali (2015) documents the British and Dutch influence on Southeast Asian legal, educational and political systems; Tan (2018) highlights the replacement of traditional Islamic curricula with Western-oriented secular education; and Khalid (1999) analyzes Russian influence on Jadidist educational reform in Central Asia. Esposito & Shahin (2016) and Mandaville (2007) emphasize the integration of Western secular ideologies and globalization in sustaining colonial imprints. Yongbao (2024) noted how Islamic law was marginalized in favor of Western civil codes. These studies collectively suggest that much structural heterogeneity within the Islamic world stems from colonial inheritance. For the latter, shared Islamic values form the second type of homogeneity, reflecting a unifying philosophical base through *ad-deen* (a way of life), a holistic framework connecting individual and state responsibilities. While cultural and political diversity exist, the integration of religion and politics creates cohesion across OIC member states. This dual identity, historical/structural and spiritual/philosophical, guides engagement with global issues, including environmental protection. Principles such as *amanah* (sacred trust), *mizan* (balance), and *maslahah* (public interest aligned with Islamic moral imperatives) remain underutilized in mainstream environmental policies.

The heterogeneity of the Islamic world, which is referred to as the MMMS, is evident in political, legal, and cultural diversity. Four typologies can be distinguished: (i) states where Islam primarily serves as cultural heritage (e.g., Albania, Benin, Guinea); (ii) states institutionalizing Islamic governance fully, including Shariah as supreme law (e.g., Afghanistan, Iran, Saudi Arabia); (iii) hybrid states applying Islamic law in specific domains while retaining secular governance elsewhere (e.g., Malaysia, Indonesia, Jordann); and (iv) secular democratic nations with cultural Islam integral to identify (e.g., Turkey, Tunisia, Senegal) (Esposito & Shahin, 2016; Maududi, 1980; Mandaville, 2007).

The third question concerns the meaning of "conservation" in the phrase "localized conservation in the Islamic world." "Conservation," in the existing discourse, may be understood as what is defined as the protection and management of natural resources and biodiversity for current and future generations (Rawat & Agarwal, 2015). Islamic conservation of the environment, however, extends beyond ecological and technical dimensions to integrate *tawhid,* ethics, law, and philosophy, framing human–nature relations morally and spiritually (Khalid, 2002).

The fourth question concerns the meaning of "localized." In this context, localization refers to the particular geographical, ecological, cultural, and institutional settings in which Muslim-majority communities live. Such diversity, whether between tropical rainforests and arid deserts, monarchies and republics, or distinct cultural traditions, necessitates conservation strategies that are adapted to local realities. For example, coral reef conservation in the Mal River cannot be directly applied to biodiversity management in Mauritania, even though both approaches are guided by the same Islamic principles. At the same time, Islam provides a universal ethical foundation for conservation. Core principles such as vicegerency—humanity's role as vicegerent or steward of the Earth, entrusted with its care and protection *(khalifah)*, the prohibition of waste, balance *(mizan)*, apply across all contexts. As Syed Muhammad Naquib al-Attas emphasized, Islam recognizes no dichotomy between the sacred and the secular; nature itself is filled with signs that point to the Divine (Al-Attas, 1995). Human interaction with the environment must therefore be grounded in tawhid and expressed through practices that are sensitive to local conditions. From this perspective, localized conservation does not weaken or compromise Islam's universal teachings. Rather, as Khalid (2002) argues, Islamic environmental ethics are not external additions but rather integral to Islam's worldview awaiting revitalization in contemporary contexts.

Methods

This study adopts a purely qualitative design, relying on secondary data and in-depth interviews with key informants. The secondary data were drawn from administrative records, literature, and policy documents. Administrative data refer to records generated by public sector institutions through their routine operations (Elias, 2014). As such, data are systematically collected for official purposes; they are considered reliable and enhance the validity of empirical research (Connelly et al., 2016). In this study, administrative records obtained from the Department of Audit served as the foundation for the case analysis presented in this chapter. The use of secondary data is well established in social science and policy research, offering advantages such as cost-effectiveness, accessibility, and the capacity to support large-scale analysis without the constraints of primary data collection (Smith, 2003; Johnston, 2014). Complementary insights were also drawn from the literature, which provided analytical perspectives and prior empirical findings, thereby strengthening the critical and comprehensive framing of the research questions. In addition to secondary sources, in-depth interviews were conducted as the primary means of data collection. A semi-structured interview format was employed, which enabled the researcher to explore informants' perspectives flexibly and context-sensitively while still adhering to a structured set of guiding questions (Creswell & Poth, 2016). A total of five government officials from different state agencies in Terengganu were selected as key informants on the basis of their institutional roles and relevance to biodiversity management in the Tasik Kenyir area. According to Hennink & Kaiser (2022), saturation can occur between five and 24 interviews, depending on purpose and heterogenity. The interviews were conducted as part of a broader research project undertaken between 2015 and 2018, supported by the Transdisciplinary Research Grant Scheme (TRGS) UMT/RMC/TRGS/15/59373(4)] under the Malaysian Ministry of Higher Education.

In the data analysis, the interview transcripts were analyzed via holistic coding, which facilitated the identification of broad, integrative themes rather than fragmenting the data into narrowly defined categories (Saldaña, 2021).

This inductive approach allowed patterns of meaning to emerge organically across diverse stakeholder perspectives, reflecting the governance, ecological, and sociocultural dimensions of conservation. To increase the robustness of the findings, the interview data were triangulated with secondary sources, including policy reports and scholarly literature. Triangulation not only deepened the analysis but also acted as a validation mechanism, reducing bias and ensuring consistency across multiple lines of evidence (Flick, 2018; Patton, 2015). The combined application of holistic coding and triangulation thus provided a rigorous framework for identifying emergent themes concerning faith, governance arrangements, and the ecological resilience of biodiversity conservation.

The geographical scope of this study involves the MMMS, together with Malaysia, with a particular emphasis on the Tasik Kenyir Geopark. Of the 57 OIC member states, 51 are classified as MMMS, spanning Africa, Asia, and Europe, yet are bound together by a shared foundation of Islamic faith. This common spiritual basis underscores the OIC's distinctive capacity to facilitate cooperation across diverse cultural, political, and economic contexts. The remaining OIC member states Guyana, Gabon, Mozambique, Suriname, Togo, and Uganda are Muslim-minority states and are excluded from the 51 MMMS considered in this study.

Tasik Kenyir Geopark conservation sites in Terengganu, Malaysia is a landscape confronting significant environmental pressures, where faith-informed governance and ecological resilience can guide sustainable conservation practices. Ecologically, Tasik Kenyir is embedded within what is widely acknowledged as the world's oldest tropical rainforest system. It is estimated to be more than 130 million years old. This ancient biogeographical formation contributed to the area's extraordinary biodiversity value. Simultaneously, the lake is a product of large-scale engineering undertaken for hydroelectric development, resulting in the impoundment of the Kenyir River and the creation of a 38,000-hectare water reservoir. Geopolitically and administratively, Tasik Kenyir lies within the Hulu Terengganu district and spans a broader land area of 209,199 hectares. This region has been gazetted under the jurisdiction of the Lembaga Kemajuan Terengganu Tengah (KETENGAH), a federal government agency. This highlights its strategic importance in regional planning and resource management. The formal designation of this area under a federal development agency reflects national-level interests in balancing ecological preservation with socioeconomic transformation, particularly through ecotourism, agroforestry, and infrastructural expansion. From a biodiversity conservation perspective, the

Tasik Kenyir region, particularly prior to 2017, was characterized by a highly contiguous forested landscape that included two permanent forest reserves resembling the Forest Reserve to the north and the Hulu Terengganu Forest Reserve to the south. The southern forest reserve shares a contiguous boundary with Taman Negara Terengganu, which itself is part of the tri-state Taman Negara complex that extends into the Pahang and Kelantan substates. This ecological corridor supports a diverse array of forest types ranging from lowland dipterocarp to hill and montane ecosystems and facilitates the migration, reproduction, and genetic exchange of flora and fauna across administrative boundaries (Sharoum et al., 2015). Since 2017, some parts of Tasik Kenyir have been developed into designated duty-free zones, consequently transforming the area into a commercial hub. The juxtaposition of ecological antiquity and modern infrastructure makes Tasik Kenyir a compelling site for transdisciplinary inquiry ranging from conservation biology to environmental governance and ecological resilience.

Results

Tasik Kenyir Case Study

The study confirms that the Tasik Kenyir Geopark reflects the persistence of a Western-centric conservation paradigm, fragmented coordination, unresolved conservation challenges, and the prioritization of economic imperatives over ecological considerations.

Forest-related biodiversity regulations applicable to the Tasik Kenyir Geopark, as shown in Table 1 in Annex A, show the dominance of the Federal List (Western-style top-down governance) and the fragmented coordination problems in Tasik Kenyir Geopark. This shows that there are 13 relevant biodiversity regulations relevant to the Tasik Kenyir Geopark. Of these, 8 (61.5%) fall under the Federal List, 3 (23.1%) under the Concurrent List, and only 1 (7.7%) under the State List. The Federal List includes major statutes such as the Wildlife Conservation Act 2010 (Act 716), the Environmental Quality Act 1974 (Act 127), and the Fisheries Act 1985 (Act 317), as well as land-use and infrastructure laws such as the Town and Country Planning Act 1976 (Act 172) and the Roads, Drains and Buildings Act 1974 (Act 133). In contrast, state-level jurisdiction is represented solely by the Terengganu Structure Plan under the State Economic Planning Unit. This concentration of authority reflects a top-down federal dominance, consistent with the Federal Constitution (Ninth Schedule), where matters under the Federal List are solely under Parliament's legislative authority. This distribution underscores the dominance of the federal government in biodiversity governance, leaving comparatively limited space for state-level decision-making. This distribution highlights how federal authority overshadows local or substate autonomy, leaving little room for substate-led and community-based conservation. Such a top-down arrangement reflects a Western-centric conservation paradigm, privileging centralized, technocratic control over context-sensitive and locally grounded approaches. Of the thirteen biodiversity-related regulations identified in Tasik Kenyir, at least light can be traced to British colonial legal origins, including the National Land Code 1965 (deriving from the 1926 Land Code) (See, 2022), the Roads, Drains and Buildings Act 1974 (from the 1937 Ordinance), the Town and Country Planning Act 1976 (modeled on the UK 1947 Act), and the Terengganu National Park Enactment 1939. Even postindependence statuses such as the Environmental Quality Act 1974 and Wildlife Conservation Act 2010 reproduce the colonial-era, Western-style governance logics. These laws institutionalized centralized authority, technocratic planning, and regulatory control while marginalizing community-based management and customary practices. The fragmented coordination across agencies in Tasik Kenyir today mirrors this legacy: overlapping jurisdictions, hierarchical decision-making, and the dominance of economic over ecological priorities are all direct continuities of colonial governance structures. This demonstrates that the current framework of biodiversity regulation is structurally inherited from British colonial administration, entrenching centralized and technocratic conservation approaches in Malaysia.

In addition, the interview findings below show that the implementation of the Tasik Kenyir Geopark policy was hindered by fragmented coordination and unresolved threats to biodiversity. The analyses of the interview findings below show that arrangements of "coordination" often take the form of parallel operations under separate mandates rather than coherent, integrated governance. For example, KETENGAH, established under the Lembaga Kemajuan Terengganu Tengah Act 1973 (Act 104), has no direct conservation authority and instead relies on initiatives such as the Kenyir Rangers, who collaborate with Fisheries, DOE, and the Marine Police. According to a representative from KETENGAH:

"Illegal hunting occurs in Tasik Kenyir; however, it falls outside the direct jurisdiction of KETENGAH. Nevertheless, KETENGAH has taken proactive measures by establishing the Kenyir Rangers team, which operates in collaboration with multiple agencies, including personnel from the Department of Fisheries. The team is also supported by the Marine Police and the Department of Environment as part of a joint effort in monitoring and patrolling activities. Additionally, the District Health Office collaborates with KETENGAH to conduct inspections of chalet areas. Should there be any issues related to hygiene, such as kitchen management or sewage systems, the Health Office holds the authority to issue fines to chalet operators who breach public health regulations. At the international level, KETENGAH's role is limited to coordination, such as participating in seminars or training programs organized by other bodies" (KETENGAH, interview, 15 June 2015).

Moreover, PERHILITAN enforces Terengganu National Park Enactment 1939 but only within gazetted park boundaries, explicitly prohibiting logging or extraction of species such as Rafflesia. An officer from the Department of Wildlife and National Parks (PERHILITAN) further asserted:

"Under the provisions of the Terengganu National Park Enactment, PERHILITAN holds jurisdiction within designated National Park areas. We have the authority to prevent any logging activities and the removal of endangered species, such as the Rafflesia flower or wild animals, as the enactment contains clauses that prohibit any biodiversity element from being extracted from the park area. Terengganu National Park has long been gazetted under the responsibility of PERHILITAN. In terms of boundary surveillance and the managementt of flora and fauna, we maintain regulatory control, as the preservation of the area falls within our jurisdiction. However, PERHILITAN's authority is confined strictly to the areas officially gazetted as part of Terengganu National Park" (PERHILITAN, interview, 10 June 2015).

In addition, the Department of Environment (DOE), empowered by the Environmental Quality Act of 1974, is responsible for regulating pollution and monitoring compliance but not for integrating watershed protection. An officer from the Department of Environment (DOE) emphasized:

"The impact on environmental quality is clearly evident. Our agency is primarily responsible for enforcement, and thus, it is imperative that we

> monitor and regulate pollution-related concerns. In such situations, enforcement activities can be undertaken. The core mandate of our agency is to ensure compliance with environmental laws and to control pollution effectively" (Department of Environment, Terengganu, interview, 28 September 2015).

An interview with the State Economic Planning Unit highlights how Tasik Kenyir's reputation as a recreational fishing site preceded its branding as an ecotourism hub, further reinforcing the tension between conservation, economic exploitation, and infrastructural purposes. An officer from the State Economic Planning Unit stated:

> "At one point, excessive fishing activities (overfishing) significantly undermined the sustainability of its fisheries. Tasik Kenyir was designated a duty-free zone following approval from the Customs Department. Tasik Kenyir plays an important role as a tourism destination and an economic generator. Historically, the lake originated as a rainwater catchment area and was initially renowned as a popular fishing site. It became widely recognized across Malaysia as a premier destination for recreational fishing" (State Economic Planning Unit, interview, 7 September 2015).

Similarly, the Department of Fisheries enforces the Fisheries Act 1985 and Amendment Act 2012, issuing fines of the RM500 for stock encroachment and conducting patrols only twice a month. An officer from the Fishery Department asserted:

> "The Department of Fisheries generally carries out enforcement against fish stock encroachment and imposes a fine of RM500 on offenders. The department also conducts patrol activities twice a month.... The Department of Wildlife and National Parks (PERHILITAN) Terengganu assisted the Tasik Kenyir Enforcement Unit in conducting operations to detect fish poisoning activities in the lake. Such practices constitute an offence under the Terengganu Inland Fisheries Enactment 1985, for which offenders may be subject to fines and imprisonment" (Fishery Department, interview, 29 September 2015).

In terms of unresolved conservation challenges, evidence from the interviews also demonstrated that significant conservation problems persist despite the regulatory abundance. Illegal hunting remains a major threat, with

KETENGAH acknowledging its prevalence but lacking jurisdiction to predict offenders (Interview, 15 June 2015). Pollution issues are evident, with DOE officers noting the "clear impacts on environmental quality" and their responsibility for enforcement (Interview, 28 September 2015). Overfishing has also undermined sustainability, exacerbatedd by Tasik Kenyir's designation as a duty-free zone a measure approved by the Customs Department to boost tourism. According to the State Economic Planning Unit, "excessive fishing activities significantly undermined sustainability of fisheries resources" (Interview, 7 September 2015). The Fisheries Department confirmed this by highlighting operations against fish poisoning under the Terengganu Inland Fisheries Enactment 1985, but enforcement capacity remains limited to fines and occasional patrols (Interview, 29 September 2015). Thus, the persistence of environmental degradation at Tasik Kenyir is not only a matter of policy implementation gaps but also a reflection of deeper structural path dependencies embedded in Malaysia's legal and institutional framework.

In addition to the aforementioned findings, the literature underscores that although biodiversity conservation initially undertaken, environmental degradation persists, with another challenge, namely, the prioritization of economic imperatives over ecological considerations, remaining prevalent. Initiatives such as the Tasik Kenyir Geopark, formerly designated National Geopark by the Malaysian National Geopark Committee on 31 May 2023, represent attempts to balance ecotourism-driven economic development with biodiversity conservation (Yusof & Jamil, 2013). Current approaches seek to localize the Sustainable Development Goals (SDGs) within three pillars social, economic, and environmental while aligning development strategies with UNESCO principles through the National Geopark Implementation Plan (PLGN) (Ministry of Natural Resources, Environment and Climate Change, 2023). The PLGN is designed to guide structured and holistic implementation, drawing on the Operational Guidelines for UNESCO Global Geoparks. However, as Mustafa et al., (2024) contend, although sustainability principles are embedded in formal policies, economic imperatives continue to be prioritized over ecological concerns, thereby compounding existing challenges. Tasik Kenyir's designation as a duty-free zone was intended to increase international tourism profile, supported by state investments of RM612.27 million in three major development projects, including the creation of a duty-free island within the lake (National Audit Department, 2023). However, the National Audit Report (2023) revealed that tourist arrivals failed to meet targets between 2017 and 2022, with the exception of 2020, which cited the incomplete development of Pulau Bay as a primary factor. The emphasis on economic priorities has produced several persistent tensions that undermine effective implementation. First, infrastructure expansion frequently conflicts with the protection of ecologically sensitive areas. Second, monitoring systems for evaluating the ecological impacts of tourism remain inadequate. Third, indigenous (Orang Asli) participation in planning and governance processes is limited, restricting opportunities for co-management and the integration of local knowledge (Rahman et al., 2023). Moreover, environmental degradation in the area continues to occur. Despite its formal recognition, Tasik Kenyir faced multiple anthropogenic pressures. Hassan et al., (2024) demonstrated that tourism and related human activities are significantly attributed to degradation, particularly through greenhouse gas (GHG) emissions and declining water quality. These stressors compromise aquatic ecosystem health and threaten local biodiversity by disrupting food chains, degrading habitats, and reducing populations of sensitive and endemic species. Environmental monitoring has further detected water pollution, while local residents have reported adverse impacts on agriculture, livestock, and overall environmental quality (National Audit Department, 2023).

As previously discussed, Malaysia, as an MMMS, faces multiple challenges in implementing biodiversity conservation strategies, as exemplified in the case of the Tasik Kenyir Geopark. Beyond the Malaysian context, the literature further indicates that similar challenges have been identified across the OIC more broadly. This study underscores these challenges as encompassing geostrategic and security considerations, postcolonial dynamics of biodiversity governance, persistent ontological and epistemological gaps, and the marginalization of indigenous knowledge systems and cosmological worldviews.

Geostrategic Importance

The biodiversity crisis faced by member states of the Organization of Islamic Cooperation (OIC) is inseparable from the wider global biodiversity emergency and has profound geostrategic and security implications. OIC regions host critical genetic, water, food, and traditional medicinal resources essential to global human well-being. The loss of biodiversity in these countries undermines not only local ecological integrity but also global stability by increasing dependency on external resources, weakening bioeconomic sovereignty, and intensifying the potential for cross-border conflicts rooted in competition over natural resources.

In recent decades, biodiversity loss worldwide has escalated to the level of a global emergency, including in areas officially designed as "protected." The Intergovernmental Science-Policy Platform on Biodiversity and Ecosystem Services (IPBES, 2019) warns that over one million species are now at risk of extinction, most within the coming decades. This is regarded as the gravest warning to date by the scientific community, reflecting systemic failures in biodiversity governance. Similarly, Leclère et al., (2020) emphasized that even with the immediate adoption of radical measures, biodiversity recovery will remain slow and feasible only in the long term. More recently, Mauterde (2025) reported that the world is unlikely to meet its biodiversity protection targets by 2030, not only because resources and capacity constraints but also because of persistent political failures and weak governance structures. Similarly, Lu (2025) reported alarming declines in global genetic diversity across more than 600 species, a trend that undermines the adaptive ability of species to disease and environmental change. These findings indicate that environmental degradation has already crossed a critical threshold that is difficult to reverse in the near term, despite ongoing international initiatives.

In this context, OIC member states face vulnerabilities. The UNEP-WCMC (2016) report highlights that biodiversity loss in the Asia–Pacific region, where many OIC states are located, is driven primarily by unsustainable development, land-use change, and environmental pollution. Kula (2001) underscores that MMMS such as Malaysia and Indonesia, both of which are recognized globally as mega biodiversity nations, are experiencing critical rates of deforestation. Moreover, nearly two dozen OIC states are water-stressed, with inefficient agricultural water use and widespread pollution from unregulated waste disposal (Butt, 2022). These dynamics severely constrain the region's capacity for climate adaptation and ecological resilience.

The strategic dimensions of biodiversity in OIC member states can be illustrated through three scenarios. First, Gulf Cooperation Council (GCC) states such as Bahrain, Qatar, and Kuwait face drastic reductions in domestic biological capacity. The ecological footprints far exceed their biocapacity, with demand for natural resources continuing to rise. This imbalance, driven by urbanization, desalination, and unsustainable exploitation, exposes them to market shocks and geopolitical conflicts over access to external resources (Qin & Imran, 2024).

Second, many OIC member states, including the MMMS, are situated in disaster-prone and politically unstable regions, where climate-related events such as droughts and floods exacerbate governance fragility. Pakistan and Iraq, for example, have experienced severe climate disasters that damaged critical infrastructure and undermined ecosystem protection (Vaghefi et al., 2015).

Third, in Southeast Asia, illegal logging, habitat fragmentation, and unsustainable exploitation have disrupted species migration corridors, especially in Malaysia and Indonesia. Woodruff (2010) documented how fragmentation accelerates the extinction of endemic species, with cascading consequences for food security and rural Muslim communities depending on forest ecosystems. With more than 51 states across Southeast Asia, North Africa, and Sub-Saharan Africa, the MMMS collectively represents some of the world's most vital ecosystems, including tropical forests, deserts, river basins, mountains, and savannahs. The decline in biodiversity within these states not only accelerated the global crisis but also heightened geopolitical and security risks. Addressing this challenge therefore requires holistic intervention, international cooperation, and collective accountability, with the MMMS of the OIC recognized as central actors in the future of global biodiversity governance.

Postcolonial Dynamics of Biodiversity Governance

The influence of the postcolonial model of biodiversity conservation governance remains evident across many OIC member states, including the MMMS. This model tends to retain Western-centric approaches that mainstream governance values rooted in colonial legacies (Salomon et al., 2023). Salomon et al., (2023) utilized the historical mainstream of biodiversity science as a tool for colonial expansion, including taxonomic classification and biodiversity research that systematically marginalized indigenous knowledge. They argue that modern science continues to be shaped by Western values such as efficiency and the reduction of ecosystems into fragmented components, without due consideration for local principles such as balance, respect, and shared responsibility.

In support of this argument, Germond-Duret (2022) maintains that contemporary conservation frameworks continue to rely on approaches that suppress indigenous communities and often ignore the realities of local livelihoods. He proposes three key dimensions of reform: the decolonization of narratives, knowledge, and practice, each aimed at dismantling the enduring colonial legacy within biodiversity governance systems. In alignment with this framework, Domínguez and Luoma (2020) highlight that conventional conservation practices, despite advances in Western science and technology, remain anchored in colonial logic. This is evident in the persistence of the "fortress conservation" model, where protected areas are established without involving local communities. They also emphasize the need to recognize local values and indigenous knowledge as integral to ecological governance reform.

In the broader context of the Global South, postcolonial political structures and weak institutional capacities often exacerbate top-down approaches to conservation. These states frequently adopt Western governance models that are misaligned with local sociocultural values, leading to the exclusion of communities from conservation processes. Political ecology theory supports this claim, as elite-led conservation tends to dismiss local knowledge systems and undervalue community-based practices (Hanna et al., 2004).

A study by Ismail et al., (2024) further illustrates that although biodiversity conservation performance has improved among low-income OIC states, there remain significant gaps in incorporating local context and participatory approaches. Member states of the OIC also face governance challenges stemming from the lack of integration of Islamic environmental ethics into mainstream conservation policies. Core Islamic epistemological concepts such as *khalifah* vicegerency), *mizan* (balance), *amanah* (trust), and *maslahah* (public interest aligned with Islamic moral imperatives) have yet to be systematically mobilized in national or international policy frameworks across most MMMS (Nasr, 2021). While contemporary Islamic scholars have increasingly recognized the significance of these principles as a foundation for environmental ethics, most MMMS have yet to assume global leadership in this discourse. As Shaleh and Islam (2024) note, Islamic environmental ethics grounded in values such as *khalifah, amanah*, and *mizan* offer a strong moral framework for ecological sustainability. Despite the growing awareness among scholars, the practical application of these principles in policymaking and global discourse remains limited in the Muslim world.

Ontological and Epistemological Gaps

One of the principal challenges in biodiversity conservation within Muslim majority states is the existence of ontological and epistemological gaps that hinder the effective utilization of local knowledge and religious values as foundational elements in the formulation of context-sensitive, sustainable and grassroots-based conservation policies. These gaps emerge when the dominant framework of modern science characterized by positivism and universalism marginalizes other forms of knowledge, including religious traditions and the Islamic worldview, from policy-making processes. This exclusion undermines both the effectiveness and local legitimacy of conservation efforts.

Ontological gaps refer to fundamental differences in assumptions about reality, for instance, questions such as "What is nature?," What is the human–nature relationship?," and "Who holds the authority to assign value to nature?" In the realm of conservation policy, these gaps manifest in the tension between divergent knowledge systems, particularly between Western modern epistemology and local knowledge systems informed by Islamic religious values. Modern Western ontology tends to conceptualize nature as a physical object that is measurable, controllable, and manageable for human benefit. The human–nature relationship is often viewed through an exploitative or utilitarian lens, with nature considered separate from both humanity and spirituality. Conversely, Islamic and local ontologies regard nature as a creation of God imbued with intrinsic value, as it manifests divine signs (āyāt). The human–nature relationship is guided by ethical and spiritual principles, wherein humans are entrusted as khalifahs (vicegerencies) responsible for maintaining cosmic balance (mizans). Nature is perceived as part of a divinely ordained system imbued with wisdom and a distinct purpose.

Epistemological gaps, on the other hand, refer to the incompatibility between differing knowledge systems, particularly between the dominant framework of modern Western science and indigenous knowledge alongside religious values. Within the context of conservation policy, such gaps become a significant barrier to the recognition, integration, and utilization of local wisdom and religious principles as legitimate sources of environmental knowledge. According to Global Biodiversity Outlook 5 (CBD, 2020), most states in the Islamic world failed to meet Aichi Biodiversity Targets 2011--2020, particularly in areas such as protected area coverage, species conservation, and local community engagement. Asia, which is home to many member states of the OIC, poses particular challenges because of its immense biodiversity coupled with high population density, making conservation goals especially difficult to attain (Farhadinia et al., 2022). The MMMS generally demonstrates imbalances in implementing conservation strategies due to multiple constraints, including institutional weaknesses, economic dependence on natural resource exploitation, and insufficient integration of religious and cultural dimensions in biodiversity policy. In addition, global biodiversity discourses largely dominated by technocratic and Cartesian epistemologies from the Global North continue to marginalize spiritual and community-rooted approaches. Major institutions such as the CBD and the Intergovernmental Science-Policy Platform on Biodiversity and Ecosystem Services (IPBES) frequently overlook Indigenous and Islamic ecological values, privileging universalized Western frameworks that reduce conservation to narrowly defined quantitative metrics (Murad, 2023). This imbalance has generated a significant epistemological gap; it not only silences alternative worldviews but also undermines conservation effectiveness by disregarding the moral, cultural, and spiritual dimensions of human–nature relations. Furthermore, this disparity reflects deeper ontological and epistemological ruptures that obstruct the inclusion of Islamic and local knowledge as foundational components of more contextualized and enduring conservation policies. Nasr (1990; 2000; 2007) famously argued that the contemporary ecological crisis arises from such ruptures within modern civilism rooted in its detachment from metaphysical principles and its failure to recognize the sacred character of the natural world.

The Marginalization of Indigenous Knowledge and Cosmological Worldviews

Another central challenge in biodiversity conservation governance is the neglect of indigenous wisdom and local cosmology. As a result, conservation frameworks often disregard the spiritual and moral bonds between communities and their environment, undermining both ecological outcomes and cultural legitimacy (Sardar, 1987). Within this context, local ecological wisdom refers to the inherited knowledge and practices transmitted across generations by local societies, including indigenous peoples and rural Muslim communities whose interactions with nature are guided by principles of balance, responsibility, and the sanctity of God's creation.

Local cosmology, meanwhile, encapsulates a worldview that links human existence with the natural world as an integrated whole, where the spiritual, social, and ecological realms are inextricably intertwined. In the Islamic tradition, this is embodied in the concept of *tawhid*, the unifying principle under the sovereignty of the Divine, and further articulated through notions including *amanah* and *khalifah*, which entrust humankind with the moral and spiritual duty of caring for the Earth. Unfortunately, modern conservation systems often operate within a secular-scientific paradigm that severs nature from spirituality and ethics, thus failing to engage local communities meaningfully and ethically.

The processes of modernization and epistemic colonization have led to the systematic erasure of these cosmological values from environmental policy and governance discourses. The colonial epistemology not only disregarded the spiritual dimension of biodiversity stewardship but also imposed a utilitarian development narrative predicated on resource exploitation. Consequently, many conservation policies fall short of their long-term goals, as they lack grounding in local communities' relational understandings of place, meaning, and environmental history.

This neglect also entrances the epistemological divide between formal knowledge (state-sanctioned science) and indigenous epistemes. In numerous cases, communities with a proven track record of sustainable stewardship are dismissed 'unscientific' simply because they do not articulate their practices in Western scientific idioms. This often results in top-down, exclusionary conservation policies that trigger conflict, resistance, and eventual policy failure.

Within the Islamic epistemological framework, indigenous ecological knowledge is not to be seen as separate from revelation or Shari'ah-based knowledge but as a dimension of *hikmah* (wisdom) grounded in *fitrah* (primordial disposition) and divine guidance. It is thus imperative to restore the role of Islamic cosmology and local ecological wisdom in the planning and implementation of conservation policies—not only to achieve sustainability but also to reassert epistemic justice and restore the dignity of marginalized communities.

By relegitimizing local cosmologies and ecological wisdom as valid sources of knowledge, we do not merely broaden the epistemological base of conservation science; we also seek a more robust moral and spiritual foundation for confronting the global environmental crisis. As Seyyed Hossein Nasr (1996) argues, the desacralization of nature under modern secular paradigms has severed humanity's spiritual relationship with the Earth, leading to ecological degradation. The revival of a sacred cosmology, particularly one rooted in Islamic metaphysics, is therefore essential for any authentic ecological ethic.

Contemporary Islamic scholars such as Izzi Dien (2000) and Nasr (1996) emphasize the importance of filtering local cosmologies through the lens of *tawhid* and the objectives of Shari'ah (maqasid al-shariah). Elements that promote environmental care, interspecies justice, and resource sustainability should be preserved and elevated, whereas components that conflict with the Islamic creed or law must be refined or discarded. This approach does not reject indigenous knowledge but rather reinforces it within a holistic Islamic value framework—thus enabling the integration of traditional wisdom, spiritual consciousness, and contemporary scientific practice in biodiversity governance. For example, the traditional ḥima conservation system practiced in Arab-Muslim societies demonstrates how Islamic cosmology and local wisdom can converge in managing natural resources sustainably. As described by Gari (2005), the ḥima system not only designates protected zones for rangelands, forests, and water sources but also emphasizes communal responsibility, trust *(amanah)*, and moral limits on the exploitation of nature.

Is local cosmology inherently at odds with Islam? The answer is far from binary. Local cosmologies are the outcome of continuous human interaction with nature and often embody values that align closely with Islamic principles such as moderation *(wasatiyyah)* and balance *(mizan)*. Nevertheless, certain elements, particularly those rooted in animistic or supersticous beliefs, are undeniably rejected, as they are prohibited in Islam. In this context, the *maqasid als-shariah* serves as an evaluative tool to ensure that endorsed values and practices contribute to the preservation of religion, life, intellect, lineage, and property. Thus, the integration of local cosmology into conservation policy becomes not only theologically legitimate but also more contextually rooted and ecologically enduring.

Discussion

As highlighted in this study, Malaysia, exemplified through the Tasik Kenyir Geopark case alongside the wider OIC region, faces multifaceted structural and institutional challenges in implementing biodiversity conservation strategies. At Tasik Kenyir, these challenges manifest as fragmented coordination, the persistence of a Western-centric conservation paradigm, unresolved ecological issues, and the prioritization of economic imperatives over ecological considerations. Meanwhile, across the OIC landscape, biodiversity loss is driven not only by ecological pressures but also by geostrategic constraints, postcolonial governance legacies, ontological and epistemological gaps, and the marginalization of indigenous knowledge and cosmological worldviews. This paradox whereby Western conservation paradigms confer symbolic legitimacy yet fail to effect substantive change.

The scholarly literature substantively supports these findings, underscoring persistent challenges in biodiversity conservation in both the Tasik Kenyir (Malaysia) context and across broader OIC landscape. Azad and Gulam (2012) observe that the predominance of Western biological conservation paradigms in Southeast Asia, including Malaysia, fails to address regional realities and overlooks Islamic ethical frameworks that could enhance conservation outcomes. In parallel, Maidin's (2012) multistakeholder study reveals that Malysia's federal constitutional structure significantly undermines biodiversity governance: states reluctance to cede land authority impedes coordination, yielding fragmented policies, funding disparities, and inconsistencies in protected area management. These insights align closely with the present study's identification of fragmented coordination as a core challenge.

Moving to the regional scale, Ismail et al., (2024) demonstrated that, paradoxically, low-income OIC member states sometimes outperform wealthy nations in biodiversity conservation between 2010 and 2020, affirming that a governance strategy, rather than economic capacity alone, fundamentally drives effectiveness. Similarly, Alam (2025), using data envelopment analysis (DEA), reported that while certain countries, such as Saudi Arabia, have prioritized investments in renewable technologies, many OIC nations continue to struggle to balance economic growth with environmental imperatives. Taken together, these studies affirm that both Malaysia and the OIC region grapple with structural, institutional, and epistemological barriers that hinder effective biodiversity conservation, even the presence of formal policy frameworks. These findings underscore the urgent need for integrative, contextually sensitive approaches grounded in Islamic ethical and faith-based governance.

Faith-based governance provides not only normative legitimacy but also tangible paths for addressing the structural and epistemological limitations observed in both local and regional conservation contexts. The existing scholarship increasingly advocates the integration of Islamic spiritual values into policy frameworks and state action as a response to the persistent challenges of biodiversity loss. Conservation efforts within OIC states hold significant potential for improvement, particularly when grounded in approaches that embrace the interconnection between Islam and nature, an ethically anchored sustainability framework, and the spiritual foundations of conservation. These dimensions offer OIC member states a unique opportunity to enhance their conservation performance and contribute more effectively to global environmental governance. Building upon the empirical and theoretical literature, the following are the recommendations for a more systematic integration of Islamic principles into environmental governance and biodiversity conservation across OIC member states, including Malaysia.

Islam–Nature Nexus in Conservation

In the context of Malaysia, Vaghefi et al., (2015) stress that Islamic values should not be treated in isolation but rather embedded within environmental policies and natural resource management frameworks. Such integration could provide a moral and spiritual anchor to otherwise technocratic governance systems. At the broader OIC level, Zagonari (2022) suggested that religion can play a pivotal role in ensuring equitable burden sharing in the pursuit of environmental sustainability, particularly in non-OECD contexts. Similarly, Al-Jayyousi et al., (2022) emphasize that morally led sustainability models grounded in ethical and spiritual traditions offer more localized, multicultural, and people-centered approaches to governance. These complement mission-led models, such as nature-based, community-based, and rights-based strategies, thereby providing multiple entry points for sustainability transitions.

From a philosophical and civilization perspective, Nasr (2007) argues that the environmental crisis is not solely technological or economic but arises from a deeper spiritual and philosophical disconnection between humanity and nature. He critiques the secular materialist paradigm for severing the sacred bond and calls for a revival of Islamic cosmology to reorient human attitudes toward the natural world. Khalid (2002) highlights the importance of Islamic ethical principles in conservation practices, whereas Khalid and O'Brien (1992) provide the theological foundation of stewardship or vicegerency *(khalifah)*, stressing that humans are trustees rather than owners of the Earth. In this view, environmental degradation constitutes a breach of divine trust and a disruption of the balance *(mizan)* ordained by God.

Thus, to operationalize the Islam–nature nexus in biodiversity conservation, several pathways are recommended. First, natural environmental policies should integrate Islamic environmental ethics such as *amanah* (sacred trust) and *mizan* (balance) into biodiversity strategies (Khalid, 2002; Vaghefi et al., 2015). This may include embedding Islamic principles in Malaysia's National Policy on Biological Diversity and OIC-wide frameworks. Second, the establishment of Islamic financing for conservation. For example, leveraging Islamic financial instruments such as *waqf* (endowments) can mobilize sustainable financing for conservation projects. Evidence shows that Islamic finance assets can reduce environmental footprints across OIC states (Rahman et al., 2023; Hassan et al., 2022). Third, in terms of faith-based environmental education, integrating Islamic ecological teachings into formal curricula, Friday sermons (*khutbah*), and community-based programs could foster public awareness and strengthen societal stewardship values (Khalid & O'Brien, 1992; Nasr, 2007). Fourth, in terms of community and indigenous participation, traditional models such as *hima* (community-conserved areas) and *harim* (protected zones) can be revitalized as culturally resonant governance tools, enhancing community can be revitalised as culturally resonant governance tools, enhancing community particularly and indigenous ecological knowledge integration (Al-Jayyousi et al., 2022). Finally, in terms of regional OIC cooperation, establishing an OIC Environmental Stewardship Framework grounded in Islamic ethics could harmonize biodiversity conservation efforts across member states while addressing shared challenges such as climate change, desertification, and biodiversity loss (Ismail et al., 2024).

Ethical Sustainability Framework

In deliberating on pathways to strengthen biodiversity conservation across OIC states, it is evident that these nations are uniquely positioned to pioneer a conservation paradigm that is simultaneously grounded in Islamic ethical principles, rooted in cultural heritage, and centered on community participation. Unlike secular models, which often treat faith and Islamic culture peripherally, OIC states can mobilize Islam's ethical sustainability framework to embed ecological stewardship within spiritual and moral responsibility. Concepts such as *khalifah* (vicegerency), *mizan* (balance), *amanah* (sacred trust), and *fasad* (prohibition of corruption and destruction) provide enduring guidance for environmental protection and resilience building (Foltz, 2003; Nasr, 1996).

Drawing on these principles, several practical policy recommendations have been proposed. First, national governments should integrate Islamic ecological ethics into their biodiversity strategies. This requires developing policy guidelines that explicitly reference the principles of *amanah* and *mizan*, ensuring that environmental protection is recognized not merely as a technical requirement but as a trust. Ministries of environment OIC states could establish dedicated Islamic Environmental Councils, composed of scholars, policymakers, and conservation experts, to ensure that biodiversity planning aligns with Shariah-informed ethical mandates. Second, Islamic finance may offer untapped resources for conservation funding. Tools such as *waqf* (endowments) can be mobilized for habitat restoration, species protection, and climate adaptation initiatives. Embedding conservation into the operational frameworks of Islamic Development Bank financing schemes could provide sustained capital for biodiversity projects while ensuring that financial mechanisms remain consistent with Islamic laws (Hassan et al., 2022). Third, the educational curricula in OIC states should mainstream Islamic ecological values, integrating lessons on vicegerency *(khalifah)* and balance *(mizan)* into science and religious studies. For example, in Indonesia, Muslim organizations such as Nahdatul Ulama (NU), Muhammadiyah, the Ulema Council of Indonesia, and pesantren-based networks actively support environmental initiatives (Mustofa et al., 2025). Programs such as Eco-Pesantren integrate Islamic education with sustainable agriculture and waste management (Herdiansyah et al., 2016). Similarly, religious leaders and institutions can leverage platforms such as *khutbahs* (Friday sermons) to disseminate ecological messages, reinforcing biodiversity protection as a spiritual obligation. This approach strengthens public environmental awareness by linking ecological responsibility with personal piety and communal accountability (Khalid, 2002). Fourth, traditional Islamic practices such as *hima* (community-conserving zones) and *harim* (protected areas) should be revitalized and adapted to contemporary biodiversity governance. These models embody principles of equitable resource use, collective responsibility, and long-term sustainability. Incorporating local indigenous knowledge and cosmologies within these frameworks can also reduce epistemological gaps in conservation policy, ensuring that governance remains both locally relevant and spiritually grounded (Al-Jayyousi et al., 2022). Finally, given the transboundary nature of ecological challenges, OIC states should establish an OIC Biodiversity and Sustainability Platform that harmonizes conservation policies across member states. Anchored in Islamic ethical principles, such a platform could facilitate joint action to address ecological threats—such as deforestation, water scarcity, and habitat loss—while also advancing the region's collective contribution to sustainable development goals. By positioning Islam at the heart of regional biodiversity strategies, OICs can provide a distinctive and normative model of conservation leadership.

Spiritual Roots of Conservation

The teachings of the Qur'an and the Prophet's s*unnah* are regarded as vital in fostering sustainable development within Islamic societies and across the globe (Bsoul et al., 2022). However, the integration of spiritual values into conservation governance represents an underutilized yet transformative pathway for biodiversity preservation in the OIC region. Historical evidence has demonstrated that faith-based ecological practices have long played a vital role in sustaining ecosystems. For example, the *hima* system, practiced for more than 1,400 years across the Arabian Peninsula, illustrates a community-based natural resource management model deeply grounded in Islamic law (Gari, 2005). By protecting grasslands, forests and water resources while engaging local communities in stewardship, *ḥima* not only preserved ecosystems but also cultivated social legitimacy for conservation policies. Similarly, in Indonesia, the governance of national parks and marine ecosystems has been framed through Islamic principles suchas *maslahah* (public interest aligned with Islamic moral imperatives), *amanah* (sacred trust), and *khalifah* (vicegerency), effectively embedding conservation within a religious and moral framework that enhances public compliance and participatory governance (Iskandar & Sofuoglu, 2025).

Thus, to operationalize the spiritual roots of conservation within the national and global biodiversity governance of OICs, member states have recommended the following: First, the government may instituteutionalise practices such as *hima* and *waqf-based* environmental endowments by embedding them within national biodiversity frameworks. Such models provide locally grounded, socially legitimate approaches that enhance conservation effectiveness while reinforcing cultural identity. Second, MMMS must actively advance Islamic principles such as *khalifah, Amanah, maslahah* and *mizan* within multilateral platforms. By foregrounding spiritual and ethical values in negotiations under the CBD, IPBES, and SDGs, OIC member states can reshape global biodiversity governance into a more inclusive and pluralistic paradigm (Mohamed, 2020). Through these principles, OICs can also reposition biodiversity conservation not only as technocratic obligations but also as moral and civilist responsibilities. In practice, this approach involves (i) introducing Islamic ethical benchmarks into global biodiversity indicators, ensuring that conservation outcomes are assessed not only by quantitative metrics but also by measures of equity, justice, and cultural relevance; (ii) advocating for the recognition of faith-based conservation models such as the *hima* system or *waqf-based* ecological endowments as legitimate frameworks within global biodiversity governance; and (iii) building South–South coalitions within the OIC to collectively propose resolutions that integrate spiritual and cultural worldviews into multilateral environmental agreements. By embedding Islamic cosmological and ethical principles into international platforms, OIC member states would not only enhance the legitimacy of conservation practices at the local level but also diversify and enrich global governance frameworks that have long been dominated by secular and technocratic paradigms. Third, the national curricula and religious platforms—such as *khutbah* (Friday sermons)—should emphasize the sacred character of nature as signs of God. This would cultivate environmental consciousness as a form of worship and moral duty, thereby enhancing grassroots commitment to biodiversity protection (Khalid, 2002). Fourth, the OIC research institutions should collaborate with indigenous communities and Islamic scholars to develop conservation methodologies that value spiritual worldviews alongside scientific models. Such epistemic and ethical pluralism strengthens ecological governance by anchoring policy in both empirical rigor and ethical-spiritual legitimacy— affirmed by Cortés-Capano et al., (2022), who advocate for recognizing multiple coexisting ethical values, rather than enforcing monolithic Western paradigms, to advance biodiversity conservation in a just and reconciling manner.

Conclusion

This chapter has examined the status of global biodiversity, with a particular focus on the challenges and opportunities for conservation within OIC member states, especially the MMMS. Through the case of the Tasik Kenyir Geopark in Malaysia, it has been shown that biodiversity governance is undermined by fragmented coordination, the persistence of Western-centric conservation paradigms, and the prioritization of economic imperatives over ecological and ethical considerations. Across the wider OIC landscape, biodiversity loss emerges not only as an ecological issue but also as a crisis of governance, justice, and epistemology where structural legacies, postcolonial trajectories, and the marginalization of indigenous and spiritual worldviews converge to limit conservation effectiveness. However, despite these constraints, significant opportunities exist. Rooted in rich Islamic civilizational traditions of stewardship, justice and balance, OIC member states are uniquely positioned to advance a conservation paradigm that is holistic, ethically grounded and context sensitive. By operationalizing Islamic principles such as *tawhid* (divine unity), *amanah* (sacred trust), *mizan* (balance), and maslahah (public interest aligned with Islamic moral imperatives), OIC member states can transcend the limitations of Western-centric models and develop frameworks that integrate the ecological, social, spiritual and cultural dimensions of sustainability.

Accordingly, several key recommendations can be made for policymakers and scholars. Policymakers should institutionalize faith-based frameworks by embedding Islamic ecological ethics into biodiversity strategies while strengthening coordination between national and local governance systems. They should also mobilize Islamic finance mechanisms such as *waqf* (endowments) and *zakat* to sustainably fund conservation initiatives and revive traditional Islamic conservation models such as *hima* (community conserved areas) and ḥarim (protected zones). In parallel, environmental ethics should be mainstreamed into education systems, religious institutions, and the public discourse to frame conservation as both a civic duty and a religious duty.

For scholars, the priority lies in developing context-specific conservation models that bridge ecological science with Islamic cosmology, producing policy-relevant knowledge that integrates faith and science. Rigorous research is also needed to evaluate the ecological impact of Islamic financial instruments and explore the epistemological synergies between Islamic stewardship principles and contemporary conservation biology. Furthermore, scholars across OIC member states should contribute to global environmental governance debates, ensuring that faith-based approaches gain recognition in multilateral frameworks such as the Convention on Biological Diversity and the Sustainable Development Goals.

In sum, biodiversity loss in the MMMS cannot be addressed solely through technocratic or economic interventions. It requires faith-informed, value-based governance that embeds Islamic ethics into biodiversity policy, financing, education, and community participation. Integrating faith, governance, and ecological resilience offers a framework that transcends purely technical models. Faith constitutes a source of knowledge of divine reality and serves as a guiding framework for moral and spiritual responsibility in matters of conservation. Ecological resilience emphasizes community acceptance, long-term sustainability, and the capacity to address biodiversity loss and climate change holistically. Together, this triad creates a context-sensitive, spiritually grounded, and socially embedded approach to conservation. By adopting such approaches, MMMS can strengthen ecological resilience and justice within their own regions while contributing meaningfully to diversifying and enriching global biodiversity governance. Islam, as a comprehensive worldview that sanctifies nature, must therefore be recognized not merely as a cultural backdrop but also as a fundamental pillar in shaping a more inclusive, just, and spiritually grounded future for global conservation.

Disclaimer

None.

References

Abdala-Roberts, L., Puentes, A., Finke, D. L., Marquis, R. J., Montserrat, M., Poelman, E. H., Rasmann, S., Sentis, A., Symons, C. C., van Dam, N. M., Wimp, G., Björkman, C., & Mooney, K. A. (2025). Connecting the dots: Managing species interaction networks to mitigate the impacts of global change. eLife, 14, e98899.

Abu-Rayash, A., & Sabbah, E. (2023). Analysis of environmental sustainability in the Holy Quran: Maqasid framework. *Journal of Contemporary Maqasid Studies, 2(1),* 61–94.

Achieng, A. O., Arhonditsis, G. B., Mandrak, N., Febria, C., Opaa, B., Coffey, T. J., Masese, F. O., Irvine, K., Ajode, Z. M., Obiero, K. and Barasa, J. E. (2023). Monitoring biodiversity loss in rapidly changing Afrotropical ecosystems: an emerging imperative for governance and research. *Philosophical Transactions of the Royal Society B, 378(1881),* 20220271.

Adams, W. M., & Hutton, J. (2007). People, parks and poverty: political ecology and biodiversity conservation. *Conservation and society, 5(2),* 147–183.

Al-Attas, S. M. N. (1978). Islam and secularism. Kuala Lumpur: ABIM.

Al-Attas, S. M. N. (1995). *Prolegomena to the metaphysics of Islam: An exposition of the fundamental elements of the worldview of Islam.* Kuala Lumpur: ISTAC.

Alam, T. (2025). Evaluating the environmental efficiency of the Organization of Islamic Cooperation in achieving climate change goals with an emphasis on Saudi Arabia. *International Journal of Innovative Research and Scientific Studies, 8(3),* 2164–2172.

Ali, M. (2015). *Islam and colonialism: Becoming modern in Indonesia and Malaya.* Edinburgh: Edinburgh University Press.

Al-Jayyousi, O., Tok, E., Saniff, S. M., Wan Hasan, W. N., Janahi, N. A., & Yesuf, A. J. (2022). Rethinking sustainable development within Islamic worldviews: a systematic literature review. *Sustainability, 14(12),* 7300.

An-Naim, A. A. (2008). *Islam and the secular state: Negotiating the future of sharia.* Harvard University Press.

Azad, M. A & Gulam, J. A. (2012). Biodiversity conservation strategy in Malaysia: A review from an Islamic perspective. *Jurnal Hadhari, 4(1),* 129–136.

Bennett, N. J., Blythe, J., White, C. S., & Campero, C. (2022). Local support for conservation is associated with perceptions of good governance, social impacts, and ecological effectiveness. *Nature Sustainability, 5(7),* 620–629.

Berkes, F. (2018). *Sacred ecology* (4th ed.). New York: Routledge.

Blankinship, L. A., Gillaspie, S., & Aboul-Enein, B. H. (2025). Highlighting the importance of biodiversity conservation through the Holy Qur'an. *Conservation Biology, 39(1),* e14309.

Brooks, T.M., Mittermeier, R.A., Da Fonseca, G.A., Gerlach, J., Hoffmann, M., Lamoreux, J.F., Mittermeier, C.G., Pilgrim, J.D. and Rodrigues, A.S. (2006). Global biodiversity conservation priorities. *Science, 313(5783),* 58–61.

Bsoul, L., Omer, A., Kucukalic, L., & Archbold, R. H. (2022). Islam's perspective on environmental sustainability: a conceptual analysis. *Social Sciences, 11(6),* 228.

Butt, M. S. (2020). State of environmental protection in the OIC states: Genesis, ongoing initiatives and recommendations from human rights perspective. *RUDN Journal of Law, 24(3),* 780–800.

Connelly, R., Playford, C. J., Gayle, V., & Dibben, C. (2016). The role of administrative datan in the big data revolution in social science research. *Social Science Research, 59,* 1–12.

Cortes-Capano, G., Hausmann, A., Di Minin, E., & Kortetmäki, T. (2022). Ethics in biodiversity conservation: The meaning and importance of pluralism. *Biological Conservation, 275,* 109759.

Creswell, J. W., & Poth, C. N. (2016). *Qualitative inquiry and design: Choosing among five approaches* (4th ed.). California: Sage Publications.

Denny, F. M. (2011). Islam and Ecology: A Bestowed Trust Holding Nature in Trust. In Tucker, M. E., & Grim, J. A. (Eds.), *Religion and Ecology: Can the climate change?* Cambridge, MA: Harvard University Press.

Díaz, S., Settele, J., Brondízio, E. S., Ngo, H. T., Agard, J., Arneth, A., ... & Zayas, C. N. (2019). Pervasive human-driven decline of life on Earth points to the need for transformative change. *Science, 366(6471),* eaax3100.

Djosetro, M & Arts, DB. (2024). How polycentric governance affects nature conservation in practice: The case of two coastal protected areas in Suriname. *International Journal of the Commons, 18(1)*, 601–617.

Domínguez, L., & Luoma, C. (2020). Decolonizing conservation policy: How colonial land and conservation ideologies persist and perpetuate indigenous injustices at the expense of the environment. *Land, 9(3),* 65.

Dryzek, J. S. (2013). *The politics of the Earth: Environmental discourses* (3rd ed.). London: Oxford University Press.

Elias, P. (2014). Administrative data. In Duşa, A, Nelle, D, Stock. G and Wagner, D, (Eds.), *Facing the future: European research infrastructures for the humanities and social sciences.* Berlin: SCIVERO, 47–48.

Esposito, J. L., & Shahin, E. E. D. (Eds.). (2016). *The Oxford handbook of Islam and politics*. Oxford University Press.

Farhadinia, M. S., Waldron, A., Kaszta, Ż., Eid, E., Hughes, A., Ambarlı, H., Al-Hikmani, H., Buuveibaatar, B., Gritsina, M. A., Haidir, I. and Islam, Z. U. (2022). Current trends suggest most Asian countries are unlikely to meet future biodiversity targets on protected areas. *Communications Biology, 5(1),* 1221.

Flick, U. (Ed.). (2024). *The SAGE handbook of qualitative research quality.* New York: SAGE Publications Limited.

Foltz, R. (2003). *Islam and ecology: A bestowed trust.* Cambridge, MA: Harvard University Press.

Germond-Duret, C. (2022). Decolonising conservation: Toward a postcolonial conservation regime. *E-International Relations.* 1–3. https://www.e-ir.info/2022/06/07/decolonising-conservation-toward-a-postcolonial-conservation-regime/.

Gari, L. (2005). *A history of the ḥima conservation system.* Cambridge: The White Horse Press.

Global Footprint Network. (2022). National footprint and biocapacity accounts 2022. Footprint Data Foundation. https://data.footprintnetwork.org/#/.

Hanna, S. S., Folke, C., & Maler, K. G. (Eds.). (1996). *Rights to nature: ecological, economic, cultural, and political principles of institutions for the environment.* Washington: Island Press.

Hassan, N. H., Rahman, N., Zawawi, M. H., & Hayana, D. (2024). Assessment of greenhouse gas emissions and water quality in relation to tourism and human activities at Kenyir Lake, Terengganu. *E3S Web of Conferences,* 589, 06005.

Hennink, M., & Kaiser, B. N. (2022). Sample sizes for saturation in qualitative research: A systematic review of empirical tests. *Social Science & Medicine, 292,* 114523.

Herdiansyah, H., Jokopitoyo, T., & Munir, A. (2016). Environmental awareness to realizing green Islamic boarding school (eco-pesantren) in Indonesia. *IOP Conference Series: Earth and Environmental Science, 30(1),* 012017.

Hussain, T. A. E., & Olawuyi, D. S. (2025). Advancing nature and biodiversity financing in the MENA region: The role of Islamic financing. In D. S. Olawuyi & R. Fakhri (Eds.), *Biodiversity and nature conservation law and policy in the Middle East and North Africa region* (pp. 206–226). Cambridge: Cambridge University Press.

IPBES. (2019). *Summary for policymakers of the global assessment report on biodiversity and ecosystem services of the Intergovernmental Science-Policy Pl Platform on Biodiversity and Ecosystem Services.* IPBES Secretariat. https://files. ipbes.net/ipbes web-prod-public-files/inline/files/ipbes_global_assessment_report_summary_for_policymakers.pdf.

IPBES. (2022). *Summary for policymakers of the methodological assessment of the diverse values and valuation of nature.* Intergovernmental Science-Policy Platform on Biodiversity and Ecosystem Services. https://zenodo.org/records/7410287.

Hussain, A. H & Brahim, M. (2006). Administrative modernization in the Malaysian local government: A study in promoting efficiency, effectiveness and productivity. Pertanika *Journal of Social Sciences & Humanities, 14(1),* 51–62.

Iskandar, H., & Sofuoğlu, H. (2025). Islamic environmentalism in Indonesia: An analytical study of MUI fatwas on environmental protection. *Bulletin of Islamic Research, 3(2),* 180–204.

Ismail, R., Abdullah, K., & Abd Ghani, L. (2024). Biodiversity conservation efforts among muslim states: Evidence from the organization of Islamic cooperation states. *Journal of International Studies (JIS), 20(1),* 83–114.

Izzi Dien, M. Y. (2000). Islamic environmental ethics, law and society. In R. C. Foltz, F. M. Denny, & A. Baharuddin (Eds.), *Environmental dimensions of Islam* (pp. 25–58). Cambridge: IUCN/Harvard University Press.

Johnston, M. P. (2014). Secondary data analysis: A method of which the time has come. *Qualitative and Quantitative Methods in Libraries, 3(3),* 619–626.

Khalid, A. (1999). *The politics of Muslim cultural reform: Jadidism in Central Asia* (Vol. 27). Univ of California Press.

Khalid, F. M. (2002). Islam and the environment. *Encyclopaedia of Global Environmental Change, 5(1),* 332–339.

Khalid, F. M, & O'Brien, J. (Eds.). (1992). *Islam and ecology*. London: Cassell.

Kula, E. (2001). Islam and environmental conservation. *Environmental Conservation. 28(1),* 1–9.

Leclère, D., Obersteiner, M., Barrett, M., Butchart, S. H. M., Chaudhary, A., De Palma, A DeClerck, F. A., Di Marco, M., Doelman, J. C., Dürauer, M. and Freeman, R. (2020). Bending the curve of terrestrial biodiversity needs an integrated strategy. *Nature, 585(7826),* 551–556.

Lu, D. (2025). *Study of more than 600 animal and plant species finds genetic diversity has declined globally.* https://www.theguardian.com/environment/2025/jan/30/study-of-more-than-600-animal-and-plant-species-finds-genetic-diversity-has-declined-globally

Maidin, A. J. (2005). *Challenges in implementing and enforcing environmental protection measures in Malaysia.* Available at SSRN: https://ssrn.com/abstract=1988124.

Mandaville, P. (2007). Globalization and the politics of religious knowledge: Pluralizing authority in the Muslim world. *Theory, Culture & Society, 24(2),* 101–115.

Maududi, S. A. (1980). *The Islamic law and constitution* (K. Ahmad, Trans.) Lahore: Islamic Publications.

Ministry of Natural Resources, Environment and Climate Change (2023). https://www.ukm.my/kelestarian/data/sdg17/pelan-pelaksanaan-geopark-negara.pdf

Mohamed, M. (2020). Islamic environmentalism and conservation ethics: Reviving the concept of khalifah. *International Journal of Ethics and Systems, 36(4),* 563–578.

Murad, M. M. (2023). The western orientation of environmentalism in the Islamic world today. *Religion and Development, 2(1),* 41–62.

Mustofa, M. L., Zenrif, M. F., & Barizi, A. (2025). Toward an Islamic ecotheology: Indonesian muslim organizations in climate mitigation and adaptation efforts. *Problemy Ekorozwoju, 20(2),* 21–31.

Nasr, S. H. (1990). *Islam and the environmental crisis*. Islamic Quarterly, 34(4), 217.

Nasr, S. H. (1996). *Religion and the order of nature.* Oxford University Press.

Nasr, S. H. (2000). The spiritual and religious dimensions of the environmental crisis. *The Ecologist, 30(1),* 18–18.

Nasr, S. H. (2007). The Contemporary Islamic Word and the Environmental Crisis. SOPHIA: *The Journal of Traditional Studies, 13(2),* 2007–2008.

National Audit Department (2023). Report of the auditor-general. https://lkan.audit.gov.my/documents/lkan/2153/CONT-KENYIR-TERENGGANU-Aktiviti-2022.pdf.

Olawuyi, D. S., & Fakhri, R. (Eds.). (2025). Case studies. In D. S. Olawuyi & R. Fakhri (Eds.), *Biodiversity and nature conservation law and policy in the Middle East and North Africa region* (pp. 87–294). Cambridge University Press.

Patton, M. Q. (2015). *Qualitative research & evaluation methods: Integrating theory and practice.* New York: SAGE Publications.

Qin, F., & Imran, A. (2024). Greening the Gulf: A deep-dive into the synergy between natural resources, institutional quality, foreign direct investment, and pathways to environmental sustainability. *Sustainability, 16(24),* 11250.

Rawat, U. S., & Agarwal, N. K. (2015). Biodiversity: Concept, threats and conservation. *Environment Conservation Journal, 16(3),* 19–28.

Saldana, J. (2021). *The coding manual for qualitative researchers.* New York: SAGE Publication.

Shaleh, A., & Islam, M. S. (2024). Averting the existential threat of the planet: Islamic environmental ethics to address the contemporary environmental crisis. *Intellectual Discourse, 32(1),* 239–264.

Sharoum, F. M, Abdullah, M. T, Ali, C. A & Ismail, R. (2015). *Geopark Tasik Kenyir.* Kuala Terengganu: Penerbit Universiti Malaysia Terengganu.

Smith, E. (2008). *Using secondary data in educational and social research.* London: McGraw-Hill Education (UK).

Sodhi, N. S. (2008). Tropical biodiversity loss and people–A brief review. *Basic and Applied Ecology, 9(2),* 93–99.

Salomon, A. K., Okamoto, D. K., Wilson, K. I. B. J., Tommy Happynook, H., Wickaninnish, Mack, W. A., ... & Augustine, A. (2023). Disrupting and diversifying the values, voices and governance principles that shape biodiversity science and management. *Philosophical Transactions of the Royal Society B, 378(1881),* 20220196.

See, A. W. L. (2022). The Torrens system in Singapore: 75 years from conception to commencement. *American Journal of Legal History, 62(1),* 66–87.

Tan, C. (2018). Colonialism, postcolonialism, Islam, and education. In Reza Arjmand & Holger Daun. *Handbook of Islamic education* (pp. 177–188). Springer.

UNEP-WCMC (2016) The state of biodiversity in Asia and the Pacific: A mid-term review of progress toward the Aichi biodiversity targets. UNEP-WCMC, Cambridge, UK. https://resources.unep-wcmc.org/products/WCMC_RT356.

Vaghefi, N., Siwar, C., & Aziz, S. A. A. G. (2015). Green economy: issues, approach and challenges in muslim states. *Theoretical Economics Letters, 5(1),* 28–35.

Valdez, J., Damasceno, G., Oh, R. R., Quintero Uribe, L. C., Barajas Barbosa, M. P., Amado, T. F., ... & Sharma, S. (2024). Strategies for advancing inclusive biodiversity research through equitable practices and collective responsibility. *Conservation Biology, 38(6),* e14325.

Visseren-Hamakers, I. J., & Kok, M. T. J. (Eds.). (2022). Transforming biodiversity governance in different contexts. In I. J. Visseren-Hamakers & M. T. J. Kok (Eds.), *Transforming biodiversity governance* (pp. 219–338). Cambridge University Press.

Woodruff, D. S. (2010). Biogeography and conservation in Southeast Asia: How 2.7 million years of repeated environmental fluctuations affect today's patterns and the future of the remaining refugial-phase biodiversity. *Biodiversity and Conservation, 19(4),* 919–941.

Wook, I. (2023). Indigenous peoples in protected areas and equitable conservation in Malaysia. *Asian Journal of Law and Policy, 3(2),* 49–64.

Yongbao, W. (2024). The impact of colonialism and nationalism on the marginalization of Islamic law in the Muslim World. *Malaysian Journal of Syariah & Law, 12(2),* 375–387.

Yuan, R., Zhang, N., & Zhang, Q. (2024). The impact of habitat loss and fragmentation on biodiversity in global protected areas. *The Science of the Total Environment, 931,* 173004.

Yusof, N. A., & Jamil, M. F. C. (2013). Exploring organizational culture dimensions in a lake-based eco-tourism resort. *GSTF Journal on Business Review, 3(1),* 1–6.

Zagonari, F. (2022). Only religious ethics can help achieve equal burden sharing of global environmental sustainability. *International Journal of Environmental Studies, 80(3),* 807–830.

Annex A.

Table 1. Forest-related biodiversity regulations applicable to the Tasik Kenyir Geopark

No.	Name of regulations	Responsible agency	Types of regulations
1	Lembaga Kemajuan Terengganu Tengah Act 1973 (Act 104)	Central Terengganu Development Authority (KETENGAH), under the Ministry of rural and Regional Development	Federal List – a federal statutory authority
2	Free Zones (Declared Area) Notification 2017) (P.U. (b) 60/2017*	Ministry of Finance; administered via the Royal Malaysian Customs Department and appointed Free Zone Authority	Federal List – Free Zones Act 1990 is federal law
3	Wildlife Conservation Act 2010 (Act 716)	Department of Wildlife and National Parks Peninsular Malaysia (PERHILITAN), under Ministry of Natural Resources, Environment and Climate Change	Federal List – matters under federal jurisdictions
4	National Forestry Act 1984 (Act 313) and Wood-Based Industries (State) Enactment 1986	Forest Department Peninsular Malaysia (Jabatan Perhutanan Semenanjung) and State Forestry Department	Concurrent List – forestry overlaps federal and state, wood-based industry for state-level
5	Roads, Drains and Buildings Act 1974 (Act 133) - Section 70A and Section 71	Department of Irrigation and Drainage (JPS), under Ministry of Environment and Water (KETTHA)	Federal List – national infrastructure and drainage control

No.	Name of regulations	Responsible agency	Types of regulations
6	Town and Country Planning Act 1976 (Act 172)	PLANMalaysia (Federal Level) and state-level Town and Country Planning Departments (State Authorities under KPKT)	Federal List – listing town and country planning as federal power; implementation often shared
7	Fisheries Act 1985 (Act 317) and Fisheries (Amendment) Act 2012 (Act A1413)	Department of Fisheries Malaysia (under Ministry of Agriculture and Food Security)	Federal List – offshore and national fisheries regulation
8	Environmental Quality Act 1974 (Act 127)	Department of Environment (JAS), under ministry of Natural Resources, Environment and Climate Change	Federal List – pollution, environmental standards
9	National Land Code 1965 (Act 56) & Lands and Mines (State Enactments)	State Land and Mines Offices and District Land Offices	Concurrent List – national land code sets framework; state administers land matters
10	Guidelines for Eco Friendly Drainage and Town Planning	PLANMalaysia, JPS and state town Departments	Federal List – drainage under federal; town planning under federal but implemented by States
11	Guidelines for Tree Cutting, Elevation, Lakes and Islands (issued under town planning system)	PLANMalaysia (Federal Town Planner) and State Town Planning Department	Concurrent List – planning area is federal-list but shared dwith State
12	State Structure Plan/Local Plan - Terengganu Structure Plan (Kenyir)	Terengganu State Economic Planning Unit (UPEN), PLANMalaysia dan Hulu Terengganu District Council	State List – land use and local planning are state responsibilities
13	Terengganu National Park Enactment 1939	PERHILITAN and State of Terengganu	Concurrent List/- national parks involve both state and federal collaboration

Source: Fieldwork 2015-2018.

Chapter 8

The Impact of Bioecological and Oceanographic Factors on Pelagic Fish Productivity in Saleh Bay, Indonesia

Aulia P. Khaerunnisa[1]
Syamsul Bahri Agus[1,*]
Taslim Arifin[2]
James P. Panjaitan[1]
Adriani[1]
Nyoman M. N. Natih[1]
Rinny Rahmania[2]
Yulius[2]
Muhammad Ramdhan[2]
Ruzkiah Asaf[2]
Hafidzuddin Fihrin[1]
Arya Ningsih[3]
Sadad[3]
and Rahmat Hidayat[3]

[1]Department of Marine Science and Technology, IPB University, Bogor, Indonesia
[2]National Research and Innovation Agency, KST Soekarno, Cibinong, West Java, Indonesia
[3]Agency for Local Research and Development, Government of Sumbawa, Indonesia

---

* Corresponding Author's Email: sba_cacul@apps.ipb.ac.id

In: Biodiversity
Editors: Mohammad Basyuni and Faujiah Nurhasanah Ritonga
ISBN: 979-8-90134-246-6

Abstract

Saleh Bay in West Nusa Tenggara Province, Indonesia, is a semi enclosed seawater that plays an important role as a habitat for small pelagic fish, which is a source of livelihood and supports the food security of coastal communities. Seasonal dynamics and changes in oceanographic conditions, such as sea surface temperature (STT), chlorophyll-a, and salinity, are known to affect distribution patterns and fish catches. This study aims to understand the relationships between these three environmental parameters and pelagic fish catches from 2020--2024. Data were obtained from the Aqua MODIS satellite, Marine Copernicus, and catch records from the Saleh Bay Fishing Port. The analysis was carried out via a combination of spatial software (ArcGIS, ODV) and cross-correlation statistical methods. The results revealed that the highest concentrations of chlorophyll-a were found during the western and early eastern seasons, indicating the high availability of natural feed during that period. SST has proven to be a dominant factor in determining the presence of fish, especially in the western season and transition periods. Cross-correlation revealed a significant negative relationship between salinity and catch ($r = -0.323$), indicating the effect of delay time on the fishery response. These findings confirm the importance of regular oceanographic monitoring as the basis for fisheries planning, which is adaptive to the seasons and changes in the marine environment, to ensure the sustainability of local stocks and food security amid climate change pressures and increased fishing activities.

Keywords: pelagic fish, chlorophyll-a, SST, salinity, fish productivity, Saleh Bay

Introduction

Pelagic fish represent a key fisheries resource in Saleh Bay, West Nusa Tenggara. These species typically form schools and migrate in response to favorable oceanographic conditions (Putri et al., 2019). Broadly, pelagic fish are classified into two groups: large pelagic fish, such as skipjack tuna (Katsuwonus pelamis), and Spanish mackerel (Scomberomorus commerson), which are commonly distributed in offshore areas (Mardiah et al., 2023), and small pelagic fish, including fringe sardines (Sardinella fimbriata), scads (Decapterus spp.), yellowstripe scads (Selaroides sp.), and anchovies (Stolephorus spp.), all of which play critical roles in marine trophic dynamics

(Nelwan et al., 2015). Oceanographic drivers such as sea surface temperature (SST), chlorophyll-a, and salinity strongly influence their distribution and population variability. SST affects metabolic processes and habitat selection (Putra, 2012), whereas chlorophyll-an indicates that phytoplankton biomass is the foundation of the marine food chain (Sulistiowati et al., 2018).

Phytoplankton serve as the primary food source for zooplankton and small pelagic fish; therefore, changes in their availability can significantly alter marine food web structures (Andriani et al., 2017). Salinity is also a determining factor affecting fish growth, survival, and feed utilization efficiency and is often considered an indicator of water productivity (Susilo et al., 2015). Extreme salinity shifts may induce physiological stress and even cause mortality. At present, oceanographic conditions can be monitored via remote sensing technologies, such as NASA's Aqua MODIS satellite, which provides spatial and temporal data on SST, chlorophyll-a, and related parameters (Sastra et al., 2018). Additionally, cross-correlation methods have been applied to examine the time lag between environmental variability and fishery production (Makridakis et al., 1999). Research conducted in East Java waters revealed that fluctuations in temperature and salinity significantly affect pelagic fish landings, particularly Sardinella lemuru (Maulidya & Wahyu, 2022). These findings emphasize the need for similar investigations in Saleh Bay to support the development of fisheries management strategies that are data driven and adaptive to oceanographic variability.

Methods

Oceanographic information, consisting of multitemporal satellite observations of sea surface temperature (SST), chlorophyll-a, and salinity, was derived from Aqua MODIS and the Copernicus Marine Service. Data on pelagic fish landings were obtained from Saleh Bay Fishing Port. Spatial patterns were analyzed via ArcGIS and Ocean Data View (ODV), whereas statistical associations were examined via the cross-correlation method (Makridakis et al., 1999).

Results

Saleh Bay (Figure 1) is located in the eastern part of Indonesia and is positioned between the Sumbawa and Dompu Regencies within West Nusa Tenggara Province (Arifin et al., 2025). The bay stretches approximately 70 km along its west–east axis and covers a marine area of approximately 1,500 km². It is identified as a semi enclosed ecosystem that remains directly linked to the Flores Sea (Julius et al., 2017).

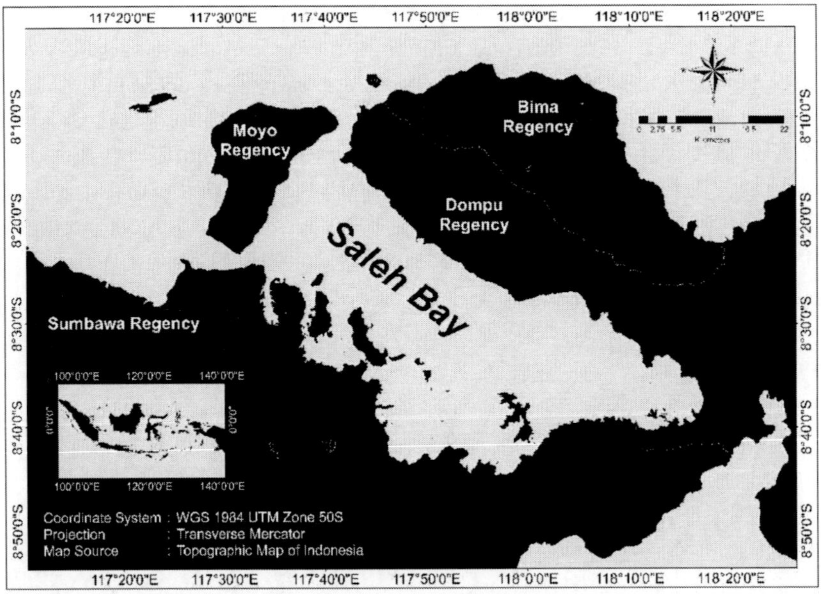

Figure 1. Map of the research location.

The fishing sector in Saleh Bay plays a crucial role in the regional economy and serves as a primary fishing ground for surrounding coastal communities (Efendi et al., 2021). The bay is characterized by distinctive oceanographic features, including mild current systems, diverse seabed morphologies, and nutrient-enriched waters that increase fishery productivity

(Arifin et al., 2025). It also harbors shallow-water ecosystems such as coral reefs, seagrass beds, and mangrove forests. Additionally, Saleh Bay is notable as a cultivation site for *Eucheuma cottonii* seaweed and as a habitat frequently visited by whale sharks (*Rhincodon typus*). The periodic occurrence of whale sharks has promoted the bay's role in marine ecotourism, offering activities such as diving, snorkeling, and direct whale shark interactions.

A seasonal evaluation of the chlorophyll-a distribution from 2020--2024 revealed a recurring annual cycle. The maximum values were recorded during the west monsoon (December–February), particularly around Dangar Rea Island and Rakit Island, which coincided with periods of increased rainfall. The concentrations decreased from the second transition season toward the east monsoon but rebounded in April and November. During the east monsoon, chlorophyll-a levels stabilized between 0.5 and 2.5 mg/m^3. These patterns indicate that seasonal variation largely governs primary productivity, with the most dynamic shifts occurring in nearshore waters and around smaller islands. Additional peaks were noted along the Sumbawa coastline in January, with another rise in March–April, especially around Rakit Island, Gili Mangrove, and Liang Island. In the first transitional season, the highest values were concentrated near the western coast of Liang Island and west of Dangar Rea before dropping from September–October (east monsoon) and rising again in November. Overall, the concentrations during both transitional phases were relatively stable, ranging from 0.5 to 1.0 mg/m^3.

Primary productivity in aquatic ecosystems is primarily controlled by light penetration, nutrient enrichment, and phytoplankton photosynthetic capacity, with chlorophyll, a commonly applied key indicator. In Saleh Bay, elevated chlorophyll-a concentrations are likely linked to riverine nutrient inflows and oceanographic processes such as currents, waves, and upwelling. Nearshore zones generally present relatively high values, particularly during the west monsoon, when rainfall increases nutrient inputs, although excessive precipitation may inhibit photosynthetic efficiency. From the east monsoon through the second transitional season, chlorophyll-a remains relatively high despite reduced rainfall, suggesting that ocean-driven mechanisms, especially upwelling, exert a dominant influence.

From 2020 to 2024, chlorophyll-a levels in Saleh Bay displayed a seasonal cycle, increasing from December, peaking around July during the east monsoon, and then declining until November (Figure 2). An exception occurred in 2024, when the highest values appeared earlier and subsequently decreased by January. As a proxy for primary productivity, chlorophyll-an indicates phytoplankton biomass, the principal food source for pelagic fish

(Fauziah et al., 2020). The increase observed during the east monsoon is attributed primarily to upwelling processes (Rosalina et al., 2024), whereas the decline corresponds to stratification driven by elevated sea surface temperatures that restrict nutrient availability (Agung et al., 2018). During the second transitional season, nutrient availability remains relatively high despite the prevailing influence of the east monsoon, thus sustaining water fertility (Rosalina et al., 2024).

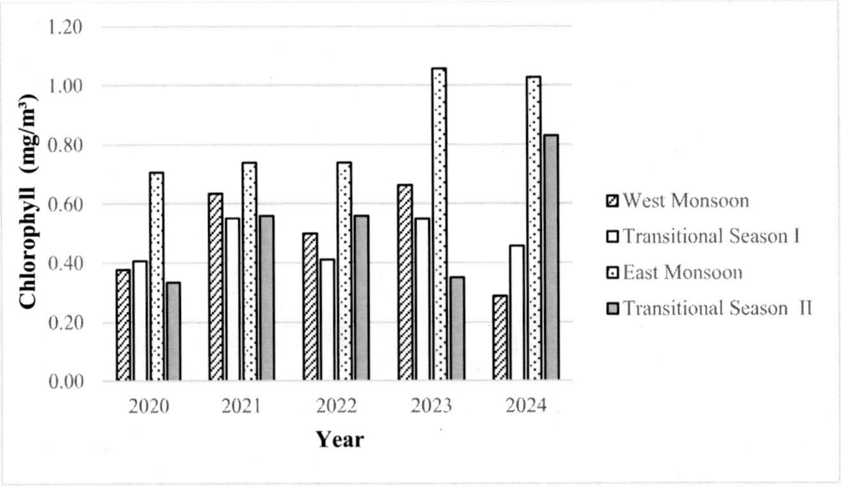

Figure 2. Variation in the average chlorophyll-a concentration in Saleh Bay from 2020-2024.

Sea surface temperature (SST) is regulated mainly by solar radiation, the strength of monsoonal winds, precipitation, and land–sea heat exchange. Shallow nearshore waters typically display higher temperatures than deeper offshore zones do. At a broader scale, SST patterns across Indonesia align with seasonal cycles: warmer conditions are observed in DJF and MAM, cooling occurs in JJA under the influence of the Australian monsoon, and temperatures rise again during SON (Napitu et al., 2015). An examination of sea surface temperature (SST) variability in Saleh Bay reveals a clear seasonal cycle. During the west monsoon (December–February) and the first transitional phases (March–May), higher temperatures are typically observed in coastal zones and estuarine areas, whereas the bay's central waters remain relatively cooler. In contrast, the east monsoon (June–August) brings cooler conditions to coastal regions because of southeast monsoon winds, whereas the inner bay stays comparatively warm. The second transitional season (September–

November) is characterized by a warming trend in the northeastern area, particularly near Liang Island and Dangar Rea Island. Interannual variations follow the same overall pattern:

1) 2020–2021: Nearshore waters warmed during the west monsoon, cooled during the east monsoon, and reheated near islands during Transition II.
2) 2022–2023: Cooling occurred along Dompu's eastern and western coasts in the west monsoon, while the central bay stayed warm into the east monsoon.
3) 2024: Elevated temperatures develop in the west-central bay during the west monsoon, followed by strong cooling in the east monsoon and then warming again in Transition II.

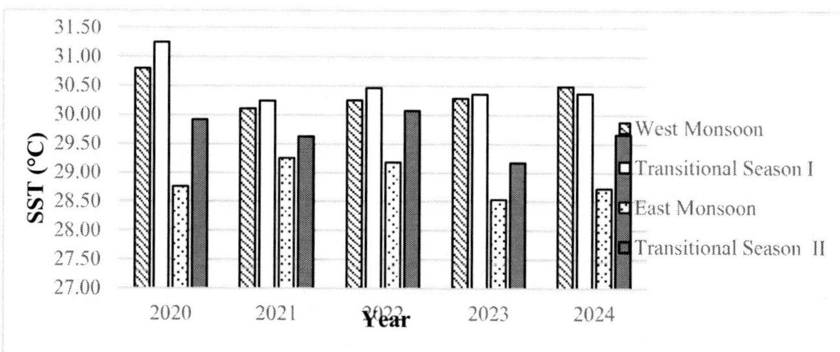

Figure 3. Seasonal average variation in SST in Saleh Bay from 2020-2024.

The mean sea surface temperature (SST) in Saleh Bay exhibited an increasing trend between 2020 and 2024 (Figure 3). In 2020, the highest values occurred before the values decreased in August. A decline was recorded from 2021–2022, followed by a rebound from 2023–2024, with peaks during December–January and decreases between July and September. SST is an essential indicator of marine environmental conditions, influencing both the metabolic activity and reproductive performance of aquatic organisms (Anggrayni et al., 2011). On a national scale, SST variations in Indonesia display a recurring seasonal cycle (Clinton et al., 2022), characterized by cooler periods during the east monsoon and the second transitional season, primarily due to southeast monsoon winds (Syafik et al., 2013). In addition, the position of the sun in the northern hemisphere during the east monsoon contributes to fluctuations in the eastern Indian Ocean (Nababan et al., 2022).

The mean sea surface temperature (SST) in Saleh Bay exhibited a clear seasonal pattern, reaching maximum levels during the west monsoon (December–February) and the first transitional period (March–May) and dropping to minimum levels during the east monsoon (June–August). The highest SST recorded was 31.6°C in March 2020, whereas the lowest, 27.9°C, occurred in September 2023. Fish catches generally increase during cooler phases, particularly in the east monsoon and second transitional seasons. SST affects both fish metabolism and upwelling processes, which transport nutrients to surface waters and create favorable feeding areas (Habib et al., 2019). In general, lower temperatures correspond with greater catches, whereas warmer waters tend to drive fish into deeper layers (Kurnianingsih et al., 2017). On an annual scale, SST variability is influenced by monsoonal wind intensity and latitudinal shifts in the sun, both of which alter circulation and heat distribution (Syafik et al., 2013).

The salinity patterns in Saleh Bay also display a regular seasonal cycle. The lowest values occur during the west monsoon (December–February) due to intense rainfall and freshwater discharge from rivers, especially along the southern and western coasts. In contrast, salinity increases during the east monsoon (June–August), with the highest concentrations observed in the central and northeastern sectors. During the first transitional season (March–May), salinity decreases until April before increasing again in May, particularly in the eastern part of the bay. The second transitional phase (September–November) shows more variability, with higher values in September–October, followed by stabilization in November, especially around island clusters. The strongest salinity gradients were observed in 2024, reflecting intensified oceanographic dynamics (Figure 4). Overall, pelagic fish were found in greater numbers at medium to high salinity levels, highlighting their importance in shaping habitat suitability and stock distribution in Saleh Bay.

Salinity conditions in Saleh Bay follow a recurring seasonal rhythm. Minimum values are observed during the west monsoon (December–February) as a result of intense rainfall and riverine discharge, especially along the southern and western coasts. In contrast, salinity increases during the east monsoon (June–August), with the highest levels recorded in the central and northeastern parts of the bay. During the first transitional phase (March–May), salinity decreases through April and then increases again in May in the eastern sector. The second transitional season (September–November) is characterized by variability, with higher levels in September–October and stabilization in November, particularly near groups of islands. The sharpest gradients were detected in 2024, indicating intensified oceanographic activity. Pelagic fish tend to be more abundant in waters with moderate to high salinity, emphasizing their role in determining habitat suitability and stock distribution in the bay.

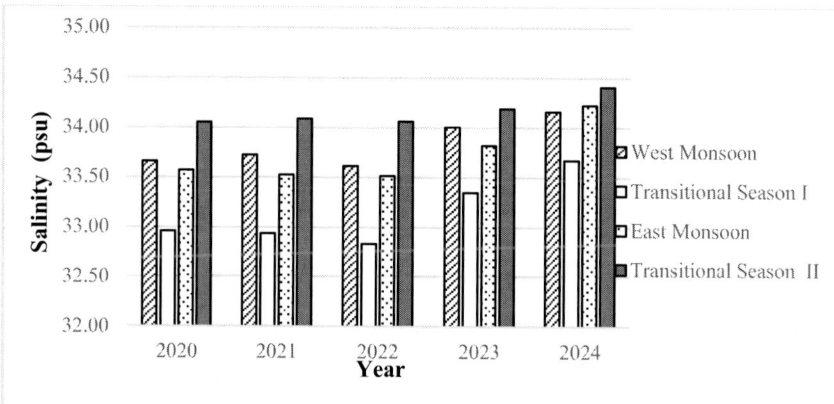

Figure 4. Variation in the average salinity season in Saleh Bay from 2020-2024.

The average salinity in Saleh Bay follows a regular yearly cycle, reaching its highest values in the second transitional season (September–October) and the lowest in the first transitional period (April–May). The maximum salinity recorded was 34.44 psu in October 2024, whereas the minimum salinity recorded, 32.83 psu, was observed in April 2020. Annual fish landings fluctuated and did not consistently correspond to salinity levels. The peak catches occurred in November 2020 (1,848,271 kg; 33.99 psu), March 2021 (17,968 kg; 33.09 psu), May 2022 (39,482 kg; 32.75 psu), December 2023 (42,488 kg; 34.05 psu), and August 2024 (42,880 kg; 34.34 psu).

These results suggest that fish tolerance to salinity occurs within an optimal range rather than at a fixed level. By means of osmoregulation, fish prefer environments where external salinity is close to their internal osmotic balance, which minimizes energy expenditure and enhances feeding efficiency as well as growth (Abidin & Ohorella, 2017). Under favorable salinity conditions, osmotic regulation requires less energy, allowing more resources to be allocated to metabolism and overall development.

Discussion

The catch per unit effort (CPUE), commonly regarded as an indicator of fishing efficiency (Listiani et al., 2017), reflects both marine productivity and environmental changes. The seasonal variation in the CPUE of flying fish in Saleh Bay shows a pattern of fluctuations between years. The year 2020 recorded a gradual increase until the peak in November. The year 2021 experienced a peak anomaly in April. In 2022, the sharp peak occurred in May and then decreased drastically. The year 2023 experienced a surge until August, allegedly due to upwelling. The 2024 CPUE increased steadily until the peak in September.

In 2020, CPUE varied substantially, from a low value of 17.88 kg/trip in August to a peak value of 1,006.68 kg/trip in November, which was largely influenced by SST and chlorophyll-a. The decline from January–August was likely linked to elevated SST, which reduced fertility (Agung et al., 2018), whereas the September–November rebound suggested more favorable conditions. In 2021, the CPUE values decreased, with a maximum of only 13.38 kg/trip (March), reflecting lower efficiency. The decrease from June–December was likely driven by monsoonal shifts that reduced productivity (Bakun, 2010). In 2022, the CPUE showed strong fluctuations in line with seasonal processes; decreases from May–October may have been caused by migration and reduced productivity, whereas increases from November–December could be tied to return migration or improved weather. Key oceanographic factors such as upwelling, SST, and salinity also contribute (Hendiarti et al., 2005). In 2023, landings rose despite stable efforts, particularly from July–October, indicating higher efficiency and stock availability, with a September peak of 327.03 kg/trip, possibly related to spawning or schooling. In 2024, the CPUE was relatively low and steady (45–55 kg/trip) from January–April but climbed between May–July (84.64 kg/trip) as the SST and productivity improved (Théophile et al., 2022), remaining high until November before declining in December. Over time, the increasing trend in CPUE may reflect enhanced fishing methods or partial recovery of stocks, although peaks could also result from temporary fish aggregations (Hoyle et al., 2023), highlighting the need for standardization in stock assessments. In general, the CPUE tends to increase from the west monsoon to the second transition, with yearly maxima occurring between April and September, underscoring the role of oceanographic processes in determining fish availability.

Chlorophyll-a does not act as a direct determinant of pelagic predator landings but exerts its influence indirectly through trophic interactions and time-delayed food-web mechanisms (Tangke et al., 2015; Bahri et al., 2017). The observed negative correlation indicates that these species are more frequently found in waters with comparatively low chlorophyll-a levels, as their primary prey are small forage fish rather than phytoplankton (Chang et al., 2022). Conversely, the positive association with SST suggests a lagged effect; nevertheless, an increase of just 1°C in SST can reduce total catches by approximately 9%, with skipjack tuna experiencing an even sharper decline of approximately 13% (Makwana & Patnaik, 2021). This can be explained by two mechanisms: (1) elevated temperatures alter feeding selectivity, and (2) energy transfer along the food chain becomes more effective in warmer waters (O'Gorman et al., 2016). Despite the relative stability of tropical SSTs, relatively high temperatures often cause pelagic fish to descend into deeper waters, thereby reducing catchability (Ekayana et al., 2017). While steady salinity favors reproduction, abrupt changes can reduce fertility and impose physiological stress on fish stocks (Chung et al., 2024). In addition, intensive pelagic fishing, particularly with large-scale or nonselective gear, may result in (a) declining mean size and biomass (FAO, 2022); (b) disruption of population structure, especially for species with slower reproductive cycles (Sadovy de Mitcheson et al., 2013); and (c) diminished recruitment due to premature harvest of juveniles (Pauly et al., 2002).

Climate influences such as ENSO, seasonal monsoon shifts, and rising sea surface temperature significantly impact pelagic fish stocks by altering productivity, migration, and reproduction. During El Niño, lower upwelling and chlorophyll-a levels in equatorial waters such as Indonesia reduce primary production and plankton availability, reducing prey for small pelagic fish and biomass (Sudradjat et al., 2024; Wei & Zhao, 2025; Fang & Zhang, 2025). Conversely, La Niña events, characterized by increased rainfall and nutrient enrichment, enhance plankton blooms and support pelagic ecosystems (Zhang et al., 2022). Terrestrial runoff lowers salinity and stratifies the water column during the west monsoon, which can limit pelagic fish development and reproduction (Chung et al., 2024). Southeast trade winds fuel the east monsoon, which upwells nutrients and increases fish catches (Maulidya & Wahyu, 2022). Fish seek favorable circumstances in deeper waters or poleward locations as the sea surface temperature (SST) increases (O'Gorman et al., 2016). Elevated SST reduces dissolved oxygen and increases metabolic demand, causing stress; disrupts reproduction in smaller taxa such as

Sardinella and Stolephorus spp. (Makwana & Patnaik, 2021); and lowers energy transfer efficiency, limiting biomass at higher trophic levels.

Conclusion

This study highlights that pelagic fish yields in Saleh Bay are strongly shaped by oceanographic parameters such as sea surface temperature (SST), chlorophyll-a, and salinity. Variations in these factors influence spatial distribution and stock availability through trophic pathways and migratory shifts. The negative link with chlorophyll-an indicates a delayed ecological response, whereas SST and salinity exert more direct effects on reproductive performance and behavior. Overall, rising temperatures and fluctuating salinities associated with climate change are expected to depress catches and pose risks to coastal food security. For effective governance, it is essential to develop a real-time monitoring framework of sea surface temperature (SST), chlorophyll-a and salinity to enable early recognition of environmental variations and resource responses. Enhancing remote sensing applications and advancing predictive models based on oceanographic indicators are equally important for guiding adaptive planning of fishing periods.

Acknowledgments

We would like to express our sincere gratitude to the Independent Campus Program (MBKM), the National Research and Innovation Agency (BRIN), and the Department of Marine Science and Technology, IPB University. Their support and collaboration have been invaluable in the successful completion of this study. The data collected in this paper are part of the thesis manuscript submitted by the first author to IPB University.

References

Abidin J, Ohorella M. 2017. The role of salinity on the survival and growth of butterflies (*Chaetodon Kleinii*). *MUNGGAI J Fisheries and Coastal Fisheries*. 3:1–7.

Agung A, Zainuri M, Wirasatriya A, Maslukah L, Subardjo P, Suryosaputro AAD, Handoyo G. 2018. Analysis of chlorophyll-a distribution and sea surface temperature as a potential fishing ground (small pelagic fish) in Kendal Waters, Central Java. *Bull Oceanography Mar*. 7(2):67–74.

Andriani, A, Damar, A, Rahardjo, M, Charles, P. H. Simanjuntak, Aries, A, Reiza, MA. 2017. Phytoplankton abundance and its role as a source of fish food in Customs Bay, West Java. *J Indopacific Aquatic Resources*. 1(2):133–144.

Anggrayni ADR, Muljo B, Jaelani LM. 2011. Study of sea surface temperature change using fashionable aqua satellites. *Geoid*. 7(1):73–78.

Arifin, T., Yulius., Ramdhan, M., Johan, O., Gunawan, D. P., Heriati, A., Mustikasari, E., Rahmania, R., Purbani, D., Dillenia, I., Agus, S. B., Asaf, R., Hatta, M., Marzuki, I., Akhwady, R., Sufyan, A., Fihrin, H., Septiningsih, E., Wahyono, A., & Athirah, A. 2025. Mariculture site selection based on water quality, cultivation, and ecological condition. *Global Journal of Environmental Science and Management*, 11(1), 129–146.

Bakun A. 2010. Linking climate to population variability in marine ecosystems characterized by non-simple dynamics: conceptual templates and schematic constructs. *J Mar Syst*. 79(3):361–373.

Chang YC, Chiang WC, Madigan DJ, Tsai FY, Chiang CL, Hsu HH, Lin SM, Zhuang MY, Sun CT, Chen LC, et al., 2022. Trophic dynamics and feeding ecology of skipjack tuna (Katsuwonus pelamis) off Eastern and Western Taiwan. *Molecules*. 27(3).

Chung M-HJ, Noble DWA, Fox RJ, Harrison LM, Jennions MD. 2024. Fluctuating salinity during development impacts fish productivity. *bioRxiv*., in press.

Clinton R, Gede Astawa Karang IW, Widiastuti W. 2022. The relationship between chlorophyll-a and sea surface temperature (SST) on the catch of lemuru sardinella lemuru fish in the Bali Strait using fashionable aqua imagery in 2009-2018. *J Mar Res Technol*. 5(1):48.

Efendi DS, Adrianto L, Yonvitner, Wardiatno Y. 2021. An Evaluation of Grouper and Snapper Fisheries Management Policy in Saleh Conference Bay, Indonesia. *Series: Earth IOP and Environmental Science*. 744(1): 1-11.

Ekayana IM, Karang IWGA, As-syakur AR, Jatmiko I, Novianto D. 2017. The relationship between tuna catch during February-March 2016 with chlorophyll-a and SST concentrations from remote sensing data in the Southern Waters of Java – Bali. *J Mar Aquat Sci*. 3(1):19.

Fang, X. & Zhang, Y. 2025. The impact of climate change and economic development on the catches of small pelagic fisheries, *Marine Policy*, 175.

Fauziah AN, Triarso I, Fitri ADP. 2020. Estimation of cod fishing areas with remote sensing technology based on chlorophyll-a parameters and sea surface temperature in Natuna Waters. *J Fish Resour Util Manag Technol*. 9 January 2019:35–44.

Habib M, Nofrizal, Mubarak. 2019. SST distribution is related to the catch of skipjack fish (Katsuwonus pelamis) in Aceh waters. *Mar Fish*. 10(1):11–22.

Hendiarti N, Suwarso, Aldrian E, Amri K, Andiastuti R, Sachoemar SI, Wahyono IB. 2005. Seasonal variation of pelagic fish catch around Java. *Oceanography*. 18(4):114–123.

Hoyle SD, Campbell RA, Ducharme-Barth ND, Grüss A, Moore BR, Thorson JT, Tremblay-Boyer L, Winker H, Zhou S, Maunder MN. 2023. Catch per unit effort modeling for stock assessment: A summary of good practices. *Fish Res*. 260(10653):1–64.

Kurnianingsih TN, Sasmito B, Prasetyo Y, Wirasatriya A. 2017. Analysis of the distribution of sea surface temperature, chlorophyll-a, and wind on the upwelling phenomenon in the Waters of Buru and Seram Islands. *J Geod Undip*. 6(1):278–287.

Kuswanto TD, Syamsuddin ML, Sunarto. 2017. The relationship between sea surface temperature and chlorophyll-a on cod fish catch in Lampung Bay. *J Fish and Marine*. VIII(2):90–102.

Listiani, Anindyas, Wijayanto, Dian, Jayanto BB. 2017. Analysis of CPUE (Catch Per Unit Effort) and the level of utilization of lemuru fishery resources (Sardinella lemuru) in the waters of the Bali Strait. *J Catch Fish J Catch Fish*. 1(1):1–9.

Makridakis S, Wheelwright SC, McGee V. 1999. *Forecasting*. 2nd edition. Erlangga, editor. Jakarta.

Makwana M, Patnaik U. 2021. Vulnerability of marine fisheries to sea surface temperature and cyclonic events: Evidence across coastal India. *Reg Stud Mar Sci*. 48(6).

Mardiah, R. S, Krisnafi, Y, Pramesthy, T. D. 2023. The large distribution of pelagic purse seine catches in Indonesia. *JSIPi (Journal of Fisheries Science and Innovation)*. 7(2):136–142.

Maulidya A, Wahyu E. 2022. The effect of oceanographic parameters on fish caught in the fishing gear of the Tancap chart in the Muncar Waters, Banyuwangi Regency. *J Sustain Agric Fish*. 1(2):96–103.

Nababan B, Sihombing EGB, Panjaitan JP. 2022. Sea surface temperature variability and chlorophyll-a concentrations in the Northeastern Indian Ocean, West Sumatra. *J Fisheries and Marine Technology*. 12(2):143–159.

Napitu AM, Gordon AL, Praise K. 2015. Intraseasonal sea surface temperature variability across the Indonesian seas. *J Clim*. 28(22):8710–8727.

Nelwan, A. F, Sudirman, M. N, Yunus, M. A. 2015. Pelagic fishing productivity in the waters of Sinjai Regency in the west–east transition season. *J Fish Sci*. 17(1):18–26.

O'Gorman EJ, Ólafsson ÓP, Demars BOL, Friberg N, Guðbergsson G, Hannesdóttir ER, Jackson MC, Johansson LS, McLaughlin ÓB, Ólafsson JS, et al., 2016. Temperature effects on fish production across a natural thermal gradient. *Glob Chang Biol*. 22(9):3206–3220.

Patty SI, Hugh R, Borrower F. 2020. Seasonal variations of temperature, salinity and surbidity of the Lembeh Strait's waters, North Sulawesi. *J Ilm PLATAX*. 8(1):110.

Putra, E. 2012. The variability of crolophyll-a concentration and sea surface temperature from fashionable satellite imagery and its relationship with pelagic fish catches in the waters of the Java Sea. IPB University.

Putri R. S, Putri A. R. S, Nurdin S, Rivai A. A. 2019. Production of large pelagic fish catch and its relationship with oceanographic parameters in fisheries management area 713, Indonesia. *J Sci Technol Util Fish Resour*. 6(11):114–127.

Romimohtarto K, Thayib SS. 1982. *Environmental and Marine Conditions in Indonesia*. Jakarta: LON-LIPI.

Rosalina D, Rizkiah R, Handayani E, Wardono S, Sutrisno BO, Ismail RM, Leilani A, Amiluddin M. 2024. Flow patterns and distribution of chlorophyll-an in the waters of the Flores Sea in 2021. *J Marine* 17(3):201–212.

Sadly M, Awaluddin A. 2017. A fish tracking system for monitoring the quality of the aquatic environment and predicting fishing locations toward sustainable fisheries management. *J Environmental Technology*. 18(1):29.

Simbolon D, Girsang HS. 2009. Study on the determination of cod fishing areas through mapping the distribution of chlorophyll-a and catch in Palabuhan Ratu, West Java. *J Indonesian Fisheries Research*. 15(4):297–306.

Sudradjat, A., Muntalif, B. S., Marasabessy, N., Mulyadi, F., Firdaus, M. I. 2024. Relationship between chlorophyll-a, rainfall, and climate phenomena in tropical archipelagic estuarine waters, *Heliyon*, 10(4).

Sulistiowati, D, Tanjung, R. H. R., Lantang, D. 2018. Diversity and abundance of plankton as a bioindicator of environmental quality in Jayapura Beach Waters. *J Biol Papua*. 8(2):79–96.

Susilo E, Islamy F, Saputra AJ, Hidayat J., Zaky A., Suniada KI. 2015. Oceanographic dynamics of the Pelagic fish catch of Kejawanan VAT from fashionable satellite data. In: *National Seminar on Remote Sensing 20015*. pp. 567–574.

Syafik A, Kunarso, Hariadi. 2013. Sea level in the Indian Ocean (Fisheries Management Area of the Republic of Indonesia 573). *J Oceanography*. 2(3):318–328.

Tangke U, Karuwal JC, Zainuddin M, Mallawa A. 2015. The distribution of sea surface temperature and chlorophyll -a affects the catch of yellowfin tuna (*Thunnus albacares*) in the waters of the Southern Halmahera Sea. *J Ipteks PSP*. 2(3):248–260.

Wei, X. & Zhao, H. 2025. Spatiotemporal distribution of chlorophyll-a concentration in the south China sea and its possible environmental regulatory mechanisms, *Marine Environmental Research*, 204.

Zhang, K., Li, M., Li, J., Sun, M., Xu, Y., Cai, Y., Chen, Z., Qiu, Y. 2022. Climate-induced small pelagic fish blooms in an overexploited marine ecosystem of the South China Sea, *Ecological Indicators*, 145.

Index